3rd International Conference on Advances in Environment Research (ICAER 2017)

IOP Conference Series: Earth and Environmental Science
Volume 68

Beijing, China
23 - 25 May 2017

Editors:

Kevin Fong-Rey Liu
Teresa Albuquerque
Eric James Strauss

ISBN: 978-1-5108-4094-2
ISSN: 1755-1307

Printed from e-media with permission by:

Curran Associates, Inc.
57 Morehouse Lane
Red Hook, NY 12571

proceedings.com

Some format issues inherent in the e-media version may also appear in this print version.

Printed by Curran Associates, Inc. (2017)

For permission requests, please contact the Institute of Physics
at the address below.

Institute of Physics
Dirac House, Temple Back
Bristol BS1 6BE UK

Phone: 44 1 17 929 7481
Fax: 44 1 17 920 0979

techtracking@iop.org

Additional copies of this publication are available from:

Curran Associates, Inc.
57 Morehouse Lane
Red Hook, NY 12571 USA
Phone: 845-758-0400
Fax: 845-758-2633
Email: curran@proceedings.com
Web: www.proceedings.com

Table of contents

Volume 68

The 3rd International Conference on Advances in Environment Research

23–25 May 2017, Beijing, China

Accepted papers received: 18 May 2017
Published online: 25 May 2017

Preface

Atmospheric Environment and Water Resources Management

IOP Conf. Series: Earth and Environmental Science **68** (2017) 011001 doi:10.1088/1755-1315/68/1/011001

PREFACE

It is our great pleasure to welcome you to 2017 3rd International Conference on Advances in Environment Research (ICAER 2017) which will be held in Jinma Hotel, Beijing, China during May 23-25, 2017. ICAER 2017 is dedicated to issues related to Advances in Environment Research.

The major goal and feature of the conference is to bring academic scientists, engineers, industry researchers together to exchange and share their experiences and research results, and discuss the practical challenges encountered and the solutions adopted. Professors from Malaysia, USA are invited to deliver keynote speeches regarding latest information in their respective expertise areas. It will be a golden opportunity for the students, researchers and engineers to interact with the experts and specialists to get their advice or consultation on technical matters, sales and marketing strategies.

This proceedings present a selection from papers submitted to the conference from universities, research institutes and industries. All of the papers were subjected to peer-review by conference committee members and international reviewers. The papers selected depended on their quality and their relevancy to the conference. The volume tends to present to the readers the recent advances in the field of Advances in Environment Research and various related areas, such as Environmental Science and Technology, Environmental dynamics, Meteorology, Hydrology, Geophysics, Atmospheric physics, Physical oceanography, Global environmental change and ecosystems management, etc..

We would like to thank all the authors who have contributed to this volume and also to the organizing committee, reviewers, speakers, chairpersons, sponsors and all the conference participants for their support to ICAER 2017.

Prof. Solomon W. Leung
Idaho State University, USA
May. 10, 2017

IOP Conf. Series: Earth and Environmental Science **68** (2017) 011002 doi:10.1088/1755-1315/68/1/011002

Conference Committee

Conference Chair

Prof. Hyo Choi, Gangneung-Wonju National University, South Korea

Prof. Ngai Weng Chan, School of Humanities, Universiti Sains Malaysia, Malaysia

Prof. Solomon W. Leung, Idaho State University, USA

Program Chair

Prof. Chan Jin Park, Incheon National University, South Korea

Prof. Jun Yao, China University of Geosciences Beijing, China

Publication Chair

Dr. Wu Jin, Beijing Jiaotong University, China

Guest Editor

Prof. Kevin Fong-Rey Liu, Department of Safety, Health and Environmental Engineering, Ming Chi University of Technology, Taiwan

Prof. Teresa Albuquerque, Instituto Politecnico de Castelo Branco, Portugal

Prof. Eric James Strauss, Michigan State University, East Lansing, USA

Technical Committee

Prof. Il-kyu Kim, Pukyong National University, South Korea

Prof. Eric James Strauss, Michigan State University, East Lansing, USA

Prof. Ahmed Jalal Khan Chowdhury, Kulliyyah (Faculty) of Science International Islamic University Malaysia

Prof. Mohd Elmuntasir Ebrahim Ahmed, Kuwait Institute for Scientific Research, Kuwait

Prof. Abbyssinia Mushunje, University of Fort Hare, South Africa

Prof. Kevin Fong-Rey Liu, Ming Chi University of Technology, Taiwan

Prof. Kei Eguchi, Fukuoka Institute of Technology, Japan

Prof. Teresa Albuquerque, Instituto Politecnico de Castelo Branco, Portugal

Prof. Ahmad Zuhairi Abdullah, School of Chemical Engineering, Universiti Sains Malaysia, Malaysia

Prof. M Ramachandra Mohan, Bangalore University, India

Prof. TALEB Mohammed Sghir, Institut Scientifique, Mohammed 5 University in Rabat, Morocco

Prof. Nour Mohammad, Premier University, Chittagong, Bangladesh

Prof. J. Helan Chandra, Jeppiaar Engineering College, Chennai, India

Prof. Qiuyan Yuan, University of Manitoba, Canada

Prof. Jabbar Hammood Al-Baidhani, Babylon University,College of Engineering, Iraq

Prof. Piyush Malaviya, Department of Environmental Science, University of Jammu, India

Prof. S S Singh, Guru Ghasidas Central University, Bilaspur, Chhattisgarh, India

Assoc. Prof. M. Azizul Moqsud, Yamaguchi University, Thailand

Assoc. Prof. Haliza Abdul Rahman, Universiti Putra Malaysia, Malaysia

Assoc. Prof. Xiang Li, National University Chiba University, Japan

Assoc. Prof. Hae Keun Yun, Department of Horticulture & Life Science, Yeungnam University, South Korea

IOP Conf. Series: Earth and Environmental Science **68** (2017) 011002 doi:10.1088/1755-1315/68/1/011002

IOP Conf. Series: Earth and Environmental Science **68** (2017) 011003 doi:10.1088/1755-1315/68/1/011003

Peer review statement

All papers published in this volume of *IOP Conference Series: Earth and Environmental Science* have been peer reviewed through processes administered by the proceedings Editors. Reviews were conducted by expert referees to the professional and scientific standards expected of a proceedings journal published by IOP Publishing.

Retrieval and Validation of Aerosol Optical Depth by using the GF-1 Remote Sensing Data

L Zhang[1,2], S Xu[3], L Wang[1,2], K Cai[1,2], and Q Ge[2,*]
[1] State Key Laboratory of Remote Sensing Science, Institute of Remote Sensing and Digital Earth, Chinese Academy of Sciences, Beijing
[2] Computer and Information Engineering College, Henan University, Henan
[3] The School of Mathematics and Computer Science, Xinyang Vocational and Technical College, Xinyang

Abstract. Based on the characteristics of GF-1 remote sensing data, the method and data processing procedure to retrieve the Aerosol Optical Depth (AOD) are developed in this study. The surface contribution over dense vegetation and urban bright target areas are respectively removed by using the dark target and deep blue algorithms. Our method is applied for the three serious polluted Beijing-Tianjin-Hebei (BTH), Yangtze River Delta (YRD) and Pearl River Delta (PRD) regions. The retrieved AOD are validated by ground-based AERONET data from Beijing, Hangzhou, Hong Kong sites. Our results show that, 1) the heavy aerosol loadings are usually distributed in high industrial emission and dense populated cities, with the AOD value near 1. 2) There is a good agreement between satellite-retrievals and in-site observations, with the coefficient factors of 0.71 (BTH), 0.55 (YRD) and 0.54(PRD). 3) The GF-1 retrieval uncertainties are mainly from the impact of cloud contamination, high surface reflectance and assumed aerosol model.

1. Introduction

Aerosol is a stable mixture of solid particles and liquid particles uniformly dispersed in the atmosphere. Aerosol Optical Depth (AOD) is the integral of the extinction coefficient in the vertical direction, which represents the attenuation of visible light by the whole atmosphere's aerosol particles. Although the aerosol content in the atmosphere is relatively small, its effect in atmospheric circulation cannot be ignored, affecting the Earth's climate and ecosystems, leading to decreased visibility on the ground, air quality deterioration, and causing a thread to public health [1]. Accurately obtain the space-time distribution, source and transmission path of aerosol, which is an important guarantee for weighting particulate pollution effect, formulating particulate reduction, prevention and control policies. Aerosol observation mainly rely on conventional instruments such as CE318 automatic tracking solar photometer and Micro Tops-II manual solar photometer, and the aerosol automatic observation network also provides important ground observation data for the research of aerosol radiation characteristics and climate effects in China [2]. However, due to the high cost, complexity of operation and maintenance of ground observation, this routine monitoring can only be carried out at limited ground stations, which cannot meet the requirement of continuous dynamic and macro monitoring of environmental pollution, and cannot reflect the spatial distribution of aerosols over a large area. Satellite remote sensing has the characteristics of large area coverage, quasi-real-time acquisition and dynamic updating. It can quickly and accurately monitor atmospheric pollution, and improve the precision of atmosphere aerosol detection [3].
GF-1 successfully launched by the Long March II Ding launch rocket on April 26, 2013 in Jiuquan Satellite Launch Center, it is the first satellite of the high-resolution Earth observation system of

IOP Conf. Series: Earth and Environmental Science **68** (2017) 012001 doi:10.1088/1755-1315/68/1/012001

national science and technology major project, it has a high space resolution (16m), repeat cycle is only 4 days, compare with the world's similar satellite, whose repeat cycle is mostly more than 10 days, GF-1 has achieve a perfect combination of high spatial resolution and high time resolution. GF-1 can provide high-precision and wide-range space observation services for the agricultural, meteorological and environmental protection departments, and play an important role in geography, oceanography and climate meteorological observation [4].

At present, with the development of research in using satellite remote sensing data to inverse land aerosol, it has formed a relatively complete system. Based on the characteristics of different load, there has been formed some aerosol inversion algorithm, such as single-channel, multi-channel, multi-angle, polarization and others. Many Chinese scholars have also conducted extensive research on aerosol satellite inversion and product application. In the study of inversion algorithms, Li et al. [5] used MODIS data to study aerosols in Beijing area, comparing the two different inversion methods of dark pixel method and structural function method. Li et al. [3] developed a surface reflectivity library algorithm and used it for inversion aerosols in urban and winter vegetation-free area. In terms of the product testing and application, Chen et al. [6] and Wu et al. [7] validate the effectiveness of MODIS aerosol optical depths in the Taiwan Strait and the Taklimakan Desert. In general, the most mature aerosol satellite products developed are based on NASA/MODIS sensors whether at home or abroad, most of the applied research and verification are also carried out for MODIS and MISR data, such as Li et al. [8] Inter-comparison of model-simulated and satellite-retrieved componential aerosol optical depths in China. At present, the research on atmospheric remote sensing for high-resolution satellite remote sensing data has just started. For example, Chinese researcher Wang Zhongting developed a land aerosol inversion algorithm based on environment-1 satellite data [9, 10]. In order to make the high-resolution satellite data play an important role in atmospheric monitoring, it is urgent to develop an aerosol optical thickness inversion algorithm based on GF-1 satellite, and evaluate its potential of inversion atmosphere parameters.

This paper mainly uses GF-1 data to develop the inversion process of AOD and validate it with ground data. The first part mainly introduces the data and processing operation used in this paper; the second part detailed elaborates the inversion algorithm and the flow; in the third part, the results are applied to Beijing-Tianjin-Hebei (BTH), Yangtze River Delta (YRD) and Pearl River Delta (PRD) where the pollution is more serious, and then compared with ground AERONET data, finally, display the conclusion.

2. Data introduction

The Ministry of Environmental Protection Satellite Environment Application Center received the original date of GF-1, those used as satellite data, including four band tiff files, rpb files, which uses to carry out geometric correction. Due to the data in 2014 is abundant and the quality is relatively stable, we extracted the original documents in 2014, through the inversion algorithm processing, then obtained the AOD results.

AERONET (Aerosol Robotic Network) is a ground-based aerosol observational network jointly established by NASA and LOA-PHOTONS (CNRS), currently there are more than 500 sites in the world, covering almost all the major regions in the world. There are about 32 observation sites in China; this paper downloaded the 2014's Level2.0 observations of all the stations. Due to the difficulties in station maintenance and complicated data correction, since 2014 in China, only Beijing (including Beijing and Xianghe stations), Hangzhou, Hangzhou and Hong Kong, the data is relatively perfect.

In order to match satellite data with ground-based observations, we have done the following processes: Based on the GF-1 transit time, we take ground observation one hour before and after to match it, obtain the average daily results. Using the AOD and Angstrom indices at 440 nm and 870 nm, obtain the AOD at 550 nm by interpolation. Based on the ground-based observation station's center latitude and longitude, extract the corresponding AOD value in satellite file within 50 km.

3. Retrieval principle and process

3.1. Dark pixel and dark blue algorithm based on GF-1 / AOD

When using the variable method for land aerosol inversion, assume the ground surface is Lambertian, atmospheric water vapor is evenly distributed, and the satellite received radiation can be expressed as [11]:

$$L(\mu_v) = L_o(\mu_v) + \frac{r}{1-rS}\mu_s F_0 T(\mu_s)T(\mu_v) \tag{1}$$

In the formula (1), $\mu_s = \cos\theta_s$, $\mu_v = \cos\theta_v$. The θ_s means the solar zenith angle, the θ_v means the observation zenith angle. The $L_0(\mu_v)$ means radiation of the obverse direction path. The r means Lambertian surface reflectivity. The S means hemispheric reflectivity of the lower atmosphere. The T means atmospheric transmittance. The μsF0 means radiation flux density, in the direction of the top of the atmosphere perpendicular to the solar. Normalization of the formula (1) can obtain the top reflectivity of the atmosphere ρ_{TOA}:

$$\rho_{TOA}(\mu_s, \mu_v, \phi) = T(\mu_s)T(\mu_v)\rho_s(\mu_s, \mu_v, \phi) / [1 - \rho_s(\mu_s, \mu_v, \phi)S] + \rho_0(\mu_s, \mu_v, \phi) \tag{2}$$

In the formula (2), the ρ_0 means equivalent reflectance of radiation in the atmosphere path. The ρ_s means surface two-dimensional reflectivity. The ϕ means relative azimuth angle. Due to the different aerosol optical thicknesses have different numerical values in S、ρ_0 and $T(\mu_s)T(\mu_v)$. Using the relationship between these variables, we can deduce the aerosol optical thickness from satellite observations, but the premise process is ground-air decoupling.

Ground-air decoupling is actually the removal of surface reflected noise. For the vegetation-intensive surface (dark pixels), because the red and blue bands have a smaller reflectivity, and there is red, blue channel reflectivity has a good linear correlation. Therefore, based on the liner relationship between red and blue bands' surface reflectivity of dark pixels, removing the surface contribution from the apparent reflectivity of the red and blue bands, then obtain the aerosol optical thickness. So, the key to aerosol optical thickness satellite remote sensing retrieval is the identification of dark pixels and the relationship between the red and blue bands [11]. Due to the lack of short-wave near-infrared channels, which is less affected by the atmosphere (Such as 2.1 microns), the GF-1 data are difficult to detect in surface information. In this paper, we select the red and near-infrared bands to construct vegetation index, then use it to identify the dark pixels. Through a large number of ground observation experiments in the North China Plain and research of related literatures [12], in this paper, the dark pixel identification threshold is set to 0.7 and the surface reflectivity relationship between red and blue bands is set to 1.6.

Usually, in urban and winter non-dense vegetation areas, it cannot use the GF-1 vegetation index to identify a wide range of dark targets and to establish an empirical relationship about reflectivity. Hsu et al. [13] proposed a dark blue algorithm; this algorithm establish the surface reflectivity library based on the blue reflectance library, which is less affected by the surface noise, and then inversion AOD of the desert and other high-level surface reflection region. This paper refer this algorithmic idea, it is assumed that there is no sudden weather effects such as rainfall and snowfall in a short time, no change in the land cover types in urban and other bright target areas, and little difference in vegetation growth. Using the GF-1 satellite transit time to get the MODIS surface reflectance data, in other words, coupling the MODIS data on the closest time as the real surface reflectivity, to solve the problem about surface reflection noise removal.

3.2. GF-1 / AOD retrieval process

In this paper, the aerosol optical thickness inversion process is shown in figure1:

3.3. Build lookup table references to preprints.

The look-up table is calculated by using 6S software for radiation transmission calculation; it needs to set different satellite observation geometrical parameters, different atmospheric aerosol parameters, considering the band of the observed data and the different surface types and other parameters. Wherein the observed geometric parameters, there are 9 solar zenith angles of 0°, 6°, 12°, 24°, 35.2°, 48°, 54°, 60° and 66°; there are 12 observation zenith angles from 0° to 66° range, and each observation angle separated by 6°; there are 16 relative azimuth angles between the suns and the satellites in the range of 0° to 180°, and each azimuth angle separated by 12°. Set up six atmospheric aerosol optical thickness values (at the wavelength of 0.55 μm): 0, 0.25, 0.5, 1.0, 1.5 and 2; the central wavelength of the band is initial set as 0.47μm, 0.66μm, and based on the band response function of GF-1 to change.

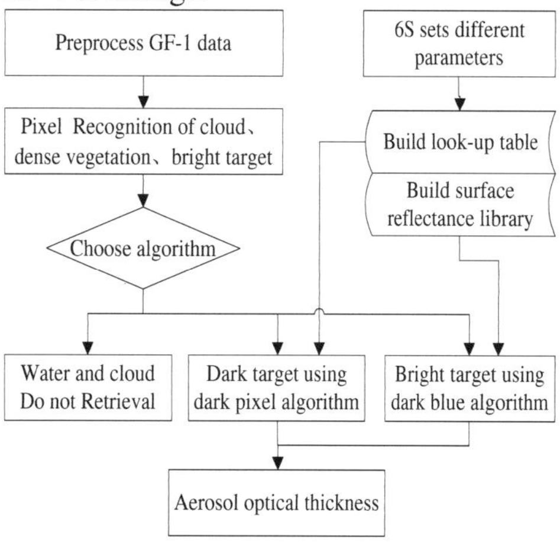

Figure 1. Aerosol optical thickness retrieval process

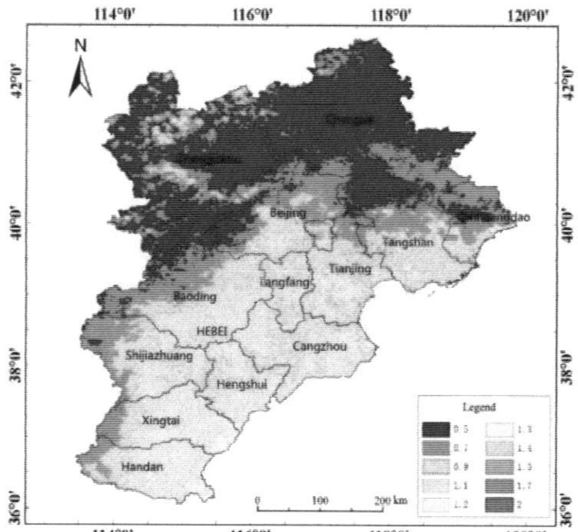

Figure 2. The annual mean distribution of aerosol optical thickness in Beijing-Tianjin-Hebei in 2014

3.4. Data preprocessing.

Data preprocessing includes calculating the apparent reflectance of 0.47μm and 0.66μm bands from the GF-1 data, simultaneously, reading geometric positioning parameters, such as latitude and longitude, altitude, solar zenith angle, solar azimuth angle, observation zenith Angle, and observation azimuth and other data.

3.5. Pixel recognition.

In this paper, the first step of the aerosol optical thickness inversion algorithm is pixel recognition. Among them, the cloud recognition algorithm is mainly based on the threshold difference of the red, blue, and near-infrared channel apparent reflectance, and compare with the MODIS surface reflectance library. As mentioned above, dense vegetation and urban bright targets and other surface pixel are mainly distinguished by vegetation indices.

3.6. Ground-air decoupling process.

According to the pixel type, the aerosol inversion is not performed in the cloud area. In the dense vegetation area, the surface contribution is removed by the surface reflectivity relation between red and blue bands. In the urban and other bright target area, through the coupling surface reflectivity library, obtain the blue band ground reflectance value of the corresponding latitude and longitude. The surface reflectance library established in this paper is obtained by using the remote sensing software ENVI, processing the 8 days synthetic surface reflectance product of MODIS by extracting band, re-projection, resampling, image cutting and mosaic.

3.7. *Calculation of aerosol optical thickness.*

Firstly, using the observed geometrical parameters read from GF-1 data and select the closest atmospheric parameters in the look-up table, then perform linear interpolation according to the actual observation geometrical data, and obtain the atmospheric parameters at different wavelengths and aerosol optical thicknesses. Finally, calculate the assumed apparent reflectivity according to formula

(2), and the real aerosol optical thickness is obtained by linear interpolation of the real apparent reflectivity.

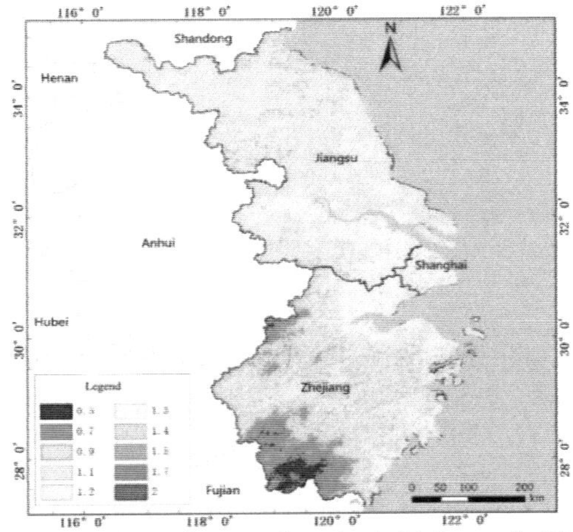

Figure 3. The annual mean distribution of aerosol optical thickness in Yangtze River Delta region in 2014

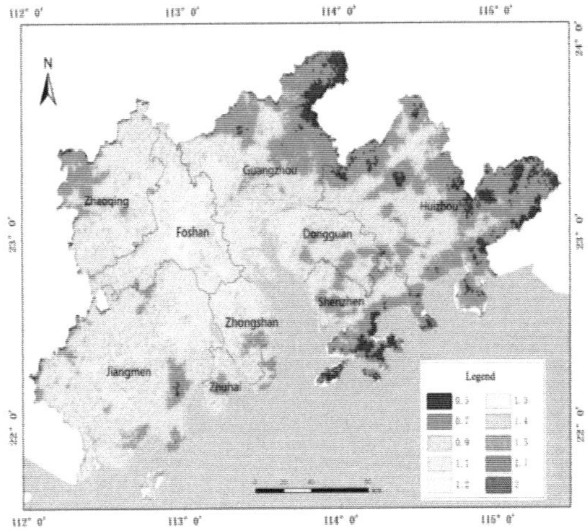

Figure 4. The annual mean distribution of aerosol optical thickness in Pearl River Delta region in 2014

4. Retrieval results and verification

4.1. Retrieval Results

With the increasingly serious air pollution situation in China, especially Beijing, Tianjin, Yangtze River Delta, Pearl River Delta and other key urban agglomerations appeared serious pollution. In this section, we monitor the aerosol distribution in the three demonstration areas based on GF-1 satellite data in 2014.

Figure 2 shows the annual mean distribution of AOD in Beijing-Tianjin-Hebei in 2014. The aerosol optical thickness shows a pattern of southeast high and northwest low. The results indicate that the distribution of AOD with higher values locate in the Beijing, Tianjin, the south of Hebei, which are core urban areas. The range is around 1. In the contrary, the areas with lower AOD values (<0.5) locate in the west and north region, which are Yanshan Mountains with few human activities and high vegetation coverage. Within the six-ring in Beijing is mainly affected by a large number of living emissions and traffic emissions. Relatively speaking, Tianjin, Hebei and other places mainly because port, industrial emissions and other air pollution.

Based on GF-1 satellite data, figure 3 shows the annual mean distribution of AOD in Yangtze River Delta region in 2014. The results indicate that the distribution of AOD with higher values (>1) are concentrated among three core cities of Shanghai, Nanjing and Hangzhou. This region include

Zhenjiang, Changzhou, Wuxi, Suzhou and other cities and towns which are intensive, manufacturing and industrial emissions of concentrated areas. In addition, Nantong in Jiangsu, Ningbo in Zhejiang, Taizhou and Wenzhou also have a high AOD distribution. Otherwise, the areas with lower AOD values locate in the north of Jiangsu, which is covered with farmland, and the southwest of Zhejiang, which is mountain with high vegetation coverage.

Figure 4 shows the annual mean distribution of AOD in Pearl River Delta region in 2014. The results indicate that the distribution of AOD with higher values locate in the core areas of Hong Kong city clusters such as Foshan, Guangzhou, Dongguan and Zhongshan, which are urban-intensive and manufacturing and industrial emissions are concentrated. With far from the above core areas, the places of AOD with lower values locate in rural and mountains with lower anthropogenic emissions and higher vegetation cover. Through the comparison results of the average annual AOD in Pearl River Delta in 2014, we can conclude that Foshan's AOD value is highest (>1.1), Zhongshan, Dongguan, Guangzhou and other cities with high level of urbanization also has a high AOD level (~1). In the contrary, the AOD value (~0.7) of Huizhou is lowest because of its highest forest cover rate.

4.2. Results verification

In this paper, we use AERONET Level 2.0 observations from Beijing, Hangzhou and Hong Kong respectively as representatives of Beijing-Tianjin-Hebei, Yangtze River Delta and Pearl River Delta, and synthesize three areas as representative of the whole country , and then to verify the inversion results of GF-1 AOD in different areas (figure 5 and table 1).

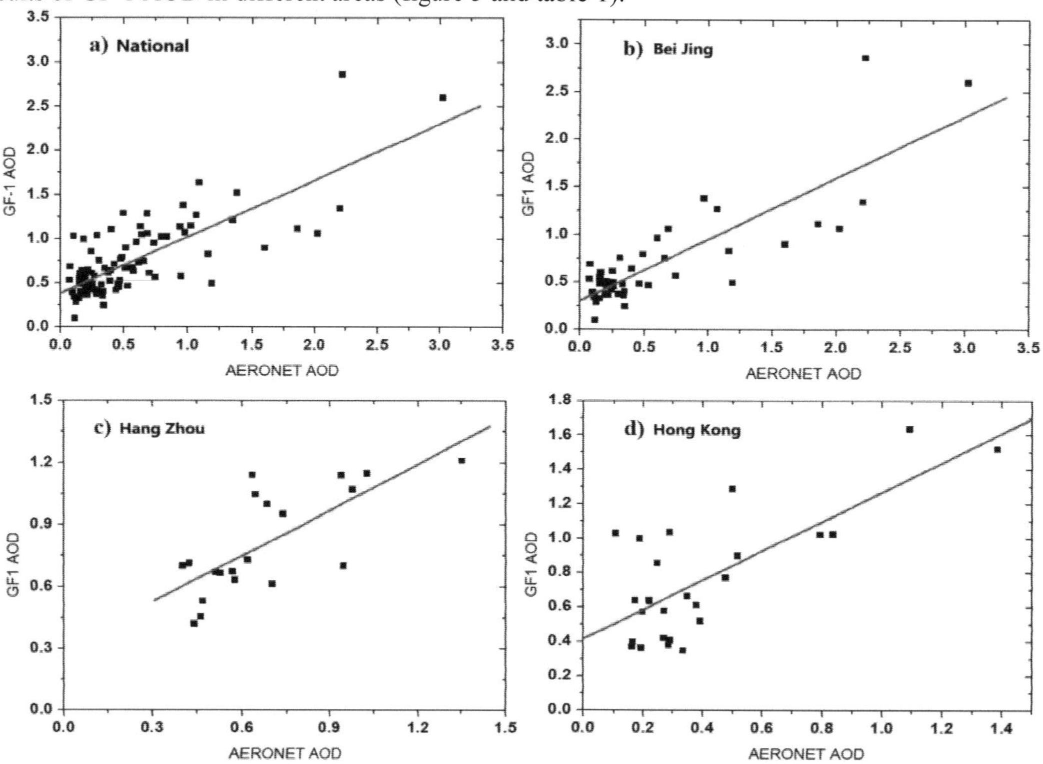

Figure 5. Comparison of scatter slots of GF-1 AOD and AERONET observations in China and three regions; (a) National (b) Beijing (c) Hangzhou (d) Hong Kong

Figure 5 and Table 1 shows the scatter slots and comparison results of GF-1 AOD and AERONET observations in China, Beijing, Hangzhou and Hong Kong in 2014 respectively. From the Table.1, the results indicate that the correlation coefficient (R2) reaches 0.6 between the ground and the satellite. The inversion range is relatively close. The MAX and MIN values differ by 0.16 and 0.03 respectively. In general, the AOD of GF-1 inversion and the ground observation data are consistent, but the satellite inversion value (0.74) is slightly higher than the observed value (0.56).

Table 1. Comparison results of GF-1 AOD and AERONET observation in China and three regions

Region	Data	N	AVG	MAX	MIN	Slope	Intercept	RMSE	R^2
National	GF-1	92	0.74	2.86	0.10	0.64	0.38	0.28	0.6
	AERONET		0.56	3.02	0.07				
Beijing	GF-1	46	0.69	2.86	0.10	0.65	0.31	0.29	0.71
	AERONET		0.60	3.02	0.07				
Hangzhou	GF-1	20	0.81	1.21	0.42	0.74	0.31	0.18	0.55
	AERONET		0.68	1.35	0.40				
Hong Kong	GF-1	26	0.76	1.64	0.35	0.85	0.42	0.25	0.54
	AERONET		0.40	1.38	0.11				

In addition, the correlation coefficient between the value of satellite inversion and AERONET in Beijing, Hangzhou and Hong Kong is 0.71, 0.55 and 0.54 respectively. The root mean square error (RMSE) is 0.29, 0.18 and 0.25 respectively. These results suggest that the three regions are well correlated. As shown in Table.1, the average AOD value of GF-1 inversion in Beijing is 0.69, the MAX is 2.86, and the MIN is 0.10. Relatively speaking, the observation of AERONET is 0.60, 3.02 and 0.07 respectively. The difference is small. However, from the slope of the scatter plot (0.65) and the intercept (0.31) point of view, there was a clear high value undervaluation situation which may be affect from the fact that Beijing is vulnerable to the factor of non-vegetation and snowfall in winter and this leads to the MODIS surface reflectivity product error becomes large. And then, it brings a greater impact to the dark blue algorithm. The results of GF-1 inversion are close to observations of AERONET in Hangzhou, and the RMSE of this area is relatively small (0.18). The analysis indicates that the inversion error of Hangzhou is mainly related to the aerosol model hypothesis and the aerosol pattern of the continental in this paper is quite different from the Yangtze River Delta.
In addition, the match sample points in Hangzhou are less, only 20. In Hong Kong, there is a significant overestimation problem with a low GF-1 inversion value (0.76) which is close to twice the ground observation (0.4). This is mainly caused by the cloud identification error of GF-1. Relative to the Beijing-Tianjin-Hebei and the Yangtze River Delta region, the Pearl River Delta cloud distribution is more. While the only four band settings of GF-1 lead to cloud recognition accuracy is significantly lower than the multi-channel MODIS and other sensors. If the thin cloud mistaken for the aerosol, it will result in significant over-estimation of the inversion results. Generally speaking, because of AERONET published less site data, this paper failed to fully verify the GF-1 results. And thus, in the future, we will consider using field test methods to obtain more ground-based data, and further the applicability of GF-1 products to test.

5. Summaries

At present, remote sensing applications using MODIS data are relatively mature, but there is little research on AOD inversion and validation based on high-resolution satellite remote sensing data. Therefore, in this paper, we use the dark pixel algorithm and the dark blue algorithm to perform the aerosol optical thickness inversion of dense vegetation and urban bright target areas respectively. And based on these results, we selected the Beijing-Tianjin-Hebei, Yangtze River Delta and the Pearl River Delta as a typical application demonstration area, the annual average distribution of the inversion results are discussed to analyze the three regions of the distribution of pollution characteristics. By using the AERONET ground-based observation data to verify the GF-1 AOD, we concluded that the correlation coefficients in national, Beijing, Hangzhou and Hong Kong were 0.6, 0.71, 0.55 and 0.54. And these indicate that GF1 inversion results can reflect the true distribution of atmospheric pollution to a certain extent. The analysis manifests that the AOD inversion error of GF1 mainly comes from the factors such as cloud recognition, surface reflectivity library accuracy and aerosol model. Therefore, these three aspects are also the focus of future AOD improvement based on high-resolution satellite data inversion.

6. Acknowledgments

This work was supported in part by National Natural Science Foundation of China under Grant 41571417.We thank the principal investigators (PI) and their staff for establishing and maintaining the AERONET sites used in this investigation. The work of Luo Zhang and Qiang Ge partially supported by the State Key Laboratory of Remote Sensing Science, Institute of Remote Sensing and Digital Earth, Chinese Academy of Sciences. We acknowledge the technical support of Shenshen Li.

7. References

[1] N. Mahowald, "Aerosol indirect effect on biogeochemical cycles and climate," *Science*, vol. 334, pp. 794-796, Nov 2011.

[2] J. Y. Xin, "Observation and research of aerosol optical properties over china by using the chinese sun hazemeter network," Ph.D. dissertation, Dept. Atmospheric science meteorology, Lanzhou Univ, 2007.

[3] S. S. Li, L. F. Chen, and J. H. Tao, "Retrieval of aerosol optical depth over bright targets in the urban areas of North China during winter," *Sci China Earth Sci*, vol. 42, no. 8, pp. 1253-1263,Aug 2012.

[4] Z. G. Bai, "Technical characteristics of GF - 1 satellite," *Aerospace China*, vol. 8, pp. 5-9, Aug 2013.

[5] X. J. Li, Y. J. Liu, H. Qiu, and Y. X. Zhang, "Retrieval method for optical thickness of aerosol over Beijing and its vicinity by using the MODIS data," *Acta Meteorologica Sinca.*, vol. 61, no. 5, pp. 580-591, Dec 2003.

[6] B. Q. Chen, and Y. M. Yang, "Remote sensing of the spatio-temporal pattern of aerosol over Taiwan Strait and its adjacent sea areas," *Acta Scientiae Circumstantiae*, vol. 28, no. 12, pp. 2597-2604, Dec 2008.

[7] X. P. Wu, J. Yang, H. Z. Che, X. J. Li, and X. A. Xia, "Verification for the satellite remote sensing products of aerosol optical depth in Taklimakan desert area," *Climatic and Environmental Research*(in Chinese), vol. 17, no. 2, pp. 149-159, Mar 2012.

[8] S. S. Li, C. Yu, and L. F. Chen et al., "Inter-comparison of model-simulated and satellite-retrieved componential aerosol optical depths in China," *Atmospheric Environment,* vol. 75, no. 141, pp: 320-332, Sep 2016

[9] Z. T. Wang, Q. Li, J. H. Tao, S. S. Li, Q. Wang, and L. F. Chen, "Monitoring of aerosol optical depth over land surface using CCD camera on HJ-1 satellite," *China Environmental Science*, vol. 29, no. 9, pp. 902-907, Sep 2009.

[10] Z. T. Wang, Q. Li, Q. Wang, S. S. Li, L. F. Chen, C. Y. Zhou, L. J. Zhang, and Y. J. Xu, "HJ-1 terrestrial aerosol data retrieval using deep algorithm," *Journal of Remote Sensing*, vol. 16, no. 3, pp. 596-610, May 2012.

[11] L. A. Remer, Y. J. Kaufman, and D. Tanré, et al., "The MODIS aerosol algorithm, products, and validation," *Journal of the atmospheric sciences*, vol. 62, pp. 947-973, Apr 2005.

[12] S. S. Li, L. F. Chen, J. H. Tao, D. Han, Z. T. Wang, and B. H. He, "Retrieval and validation of the surface reflectance using HJ-1-CCD data," *Spectroscopy and Spectral Analysis*, vol.31, no. 02, pp. 516– 520, Feb 2011.

[13] N. C. Hsu, S. C. Tsay, and M. D. King, et al., "Aerosol properties over bright-reflectance source regions," *IEEE T. Geosci. Remote*, vol. 42, pp. 557-569, 2004.

Carbon Dioxide Emissions Due to Forest Fires in Bukit Batu Area, Bengkalis Regency, Indonesia

Sofia Anita[1]; T Ariful Amri[2], T Abu Hanifah[3], Edo Furnando[4], Amos Lukas[5]

[1] Sofia Anita is lecturer at Chemistry Dept. University of Riau
[2]T Ariful Amri is lecturer at Chemistry Dept. University of Riau
[3]T Abu Hanifah is lecturer at Chemistry Dept. University of Riau
[4]Edo Furnando was the University of Riau Student
[5]Amos Lukas is staff of Agency for the Assessment and Application of Technology, Jakarta

Abstract. High concentration of carbon dioxide in the atmosphere is the major cause of global warming. This study focuses on estimation of carbon emissions from forest fires in Indonesia, especially Bukit Batu area, Bengkalis Regency. Peatlands in this area are widely used as an agricultural cultivation and plantations. The aim of this study is to measure the concentration of CO_2 emitted based on the relationship of physical and chemical properties of peat soil. Measurements carried out on these peatlands with different vegetation covered, i.e. bush land, palm plantations and secondary forests. Methods used in this research were Infrared Gas Analyzer and Gas Chromatography. The average of CO_2 emissions obtained of bush land, palm plantations, and secondary forest were 497.4 ppm; 523. 2 ppm; and 457.2 ppm, respectively.

1. Introduction

Indonesia has total land area in about 188.2 million hectares, consists of 148 million hectares of dry land and 40.2 million hectares of wetlands. Approximately 20.6 million hectares of the total wetlands are peatlands [1]. Riau is a province that has the largest peat land in Sumatra reached of 4,360,740 hectares and 803,891 hectares in Bengkalis Regency [2]. Peatland is one form of a wetland as a substrate of acidic and generally marginal lands for agricultural purposes and plantations. Nonetheless, in the last two decades there has converted to the peatland for agriculture and plantations on a large scale.

Converted peatlands have usually preceded by deforestation or clearing of forest and burning the land. Then, followed by drainage or excavation of canals and ditches those lead to wetland loss of substantially water. Therefore, the water level in the peat that has converted to shrinking so that peatlands are no longer stagnant and becomes more porous surface layer and aerobic [3]. This condition can lead to the decomposition of peat [4]-[5]. As a deposit of organic material, peat has a very large carbon content. Every one meter of peat layers expects to store around 700 tons Carbon yr^{-1} ha^{-1}, when it decomposed releases carbon in the form of CO_2 gases [5]. Carbon dioxide is one of the most important greenhouse gas, which contributed to global climate changes [6].

The release of CO_2 from peat land as palm plantations, bush land and secondary forest were done by [7] in the Province of Central Kalimantan and West Kalimantan. This research conducted in the Riau Province, especially in Bukit Batu area, Bengkalis Regency. Three type of peatlands (palm plantations, bush land and secondary forest) analyzed using Gas Chromatography equipped with Thermal Conductivity Detector and Gas Chromatography equipped with Thermal Conductivity Detector (Porapack Q column i.d. 2 m × 3 mm film thickness). As a carrier gas was used H_2 with flow rate of 50 mL min^{-1}, and Infrared Gas Analyzer instrument.

IOP Conf. Series: Earth and Environmental Science **68** (2017) 012002 doi:10.1088/1755-1315/68/1/012002

2. Experimental method

Equipment used in this research were a chamber made of PVC (diameter 280 mm and height 460 mm, the data recorder (VR-71, T & D Japan Co.), stopwatch, GPS (Global Positioning System), thermometer, batteries (GM 7Z-4A 12 V-8AH), 30 mL glass vial, Gas Chromatography with Thermal Conductivity Detector (GC-TCD) Shimadzu 2014, Infrared Gas Analyzer (GMT221 Vaisala, Finland). Materials used were hydrogen gas (H_2) as a carrier gas at Gas Chromatography with Thermal Conductivity Detector (GC-TCD) Shimadzu, 2014.

Measurements and sampling analysis include: (1) site selection; (2) the installation of the chamber; (3) measurement of physical and chemical properties of peat that covers the depth of the ground water level, peat temperatures, and pH (H_2O); (4) measurement of CO_2 flux. Sampling location was Bukit Batu areas, Bengkalis Regency. The coordinate was 35°N, 136°E). Based on physiographic types of land, the type of land that are bush land, palm plantations and secondary forests. The different characteristics of each property was determined based on the physical and chemical conditions such as depth of water table peat soil, peat temperature and pH (H_2O).

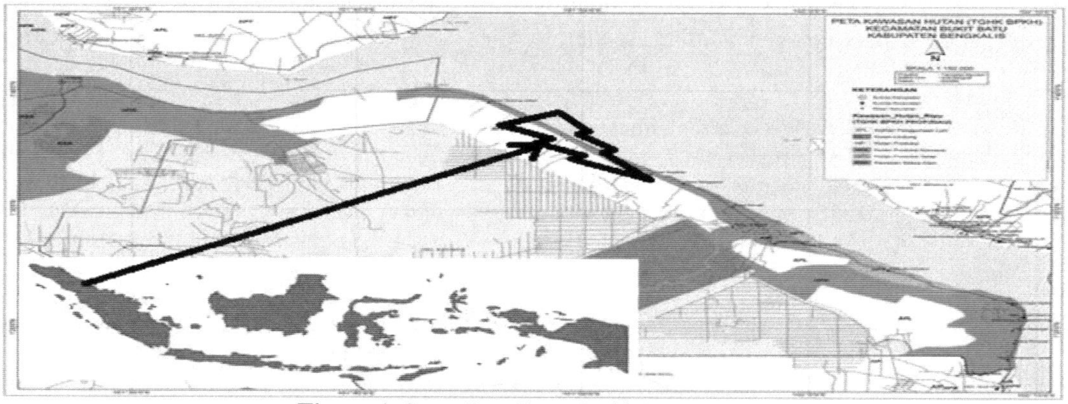

Figure 1. Sampling Sites (Bukit Batu Area)

Measurement of CO_2 flux in this area determined by two methods: Infrared Gas Analyzer and Gas Chromatography. Bush land samples and Secondary Forest analyzed using Infrared Gas Analyzer method. Samples measured through several points based on the distance between the main drainage channels with a sampling point. The placement point determined by making transects using Global Positioning System (GPS). The type of the selected transect is a straight line drawn from 5 m, 25 m, and 40 m to the main drainage channels in those area. While, GC method used only for palm plantation samples. Chamber used cylinder made of PVC with a diameter of 280 mm and height of 460 mm. Three chambers installed in the form of a triangular formation on the surface of the land for each sampling point. Each chamber has many holes as an entry of the roots of plants. Capturing and measuring CO_2 flux one day after the installation of the chamber. The chamber also was equipped with a slit septum for a syringe fitted to the bottom of the chamber. The flux of CO_2 trapped in the chamber took as many as 30 mL and then transferred into a glass vial. The sampling time interval starts at 0.3 and 6 minutes after the lid fitted [4]. CO_2 flux calculated using the following formula:

$$fc = \frac{Ph}{RT} x \frac{dC}{dt} \tag{1}$$

where, fc = CO_2 flux (μmol $m^{-2}sec^{-1}$); P = pressure based on IR data GA (Pa), h = chamber height R = gas constant (8,314 Pa m^3/K/mol); T = temperature (K); dC/dt = CO_2 concentration change [8].

Gas samples analyzed by GC-TCD at Kyoto University, Japan. Gas Chromatography equipped with Thermal Conductivity Detector (Porapack Q column i.d. 2 m x 3 mm film thickness. Hydrogen used as a carrier gas with flow rate of 50 mL min^{-1}.

3. Results and discussions

The amount of CO_2 emitted flux of peat is highly dependent on the characteristics of the peat, and therefore the need to analyze the condition of the turf.

The results of measurement of the condition of the peat covering the physical and chemical properties of peat soil in the form of temperature, pH (H_2O), and the depth of the ground water level shown in Table 1.

3.1. CO2 flux

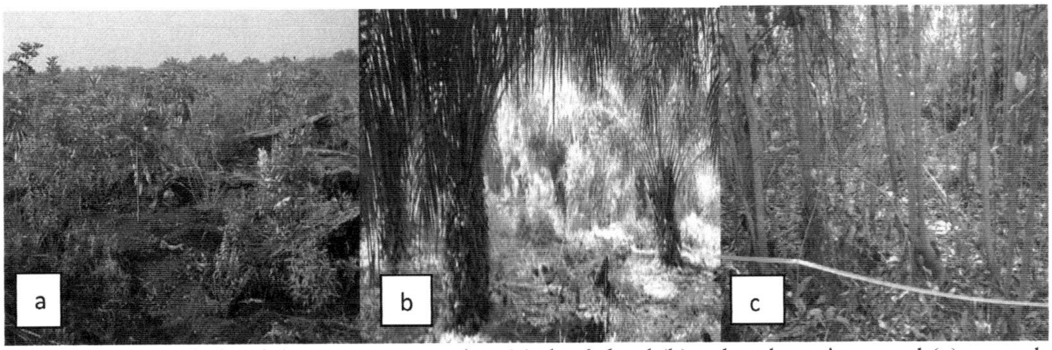

Figure 2. Characteristic of three type vegetation (a); bush land (b) palm plantations; and (c) secondary forest.

Table 1. Average Temperature of Soil, pH (H_2O), and Depth of Water Table

Land Types	Soil Temperatures (°C)	pH (H_2O)	Depth of Water Table (cm)
Bush Land	**31,1±0,77**	6,3±0,51	37,0±22,07
Palm Plantation	29,7±1,44	**6,0±0,46**	46,7±37,39
Secondary Forest	24,2±0,63	6,4±0,44	**77,7±16,66**

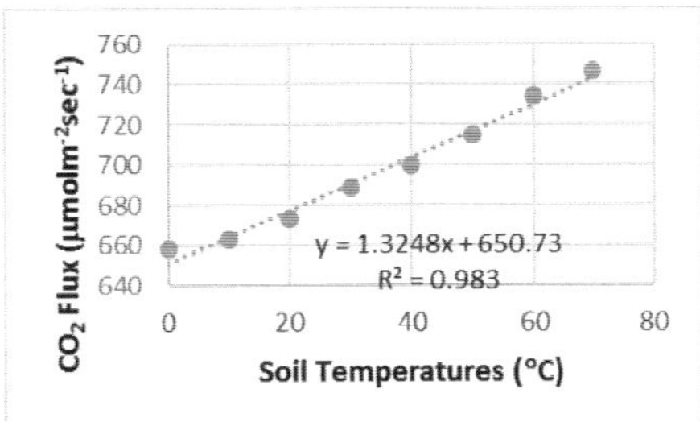

Figure 3. Relationship between soil temperatures and CO_2 Flux

Relationship between soil temperatures and CO_2 Flux showed in Figure 3. It shows strong positive correlation (R = 0.983). An average measurement of each type of sampling sites toward soil temperature presented on Figure 4. There were 497.4 ppm CO_2 flux (bush land), 523.2 ppm (palm plantation), and 457.2 ppm (secondary forest). Furthermore, Figure 5 and Figure 6 show an average of pH (H_2O), and the depth of water table on each location.

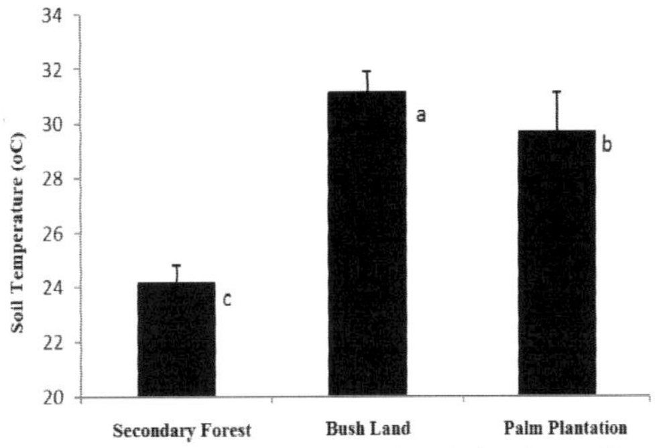

Figure 4. Comparison of soil temperature on peat soil of bush land (n = 9), palm plantations (n = 9), and secondary forest (n = 9). Sign different letters (a, b, c) indicate significant differences at $\alpha = 0.05$.

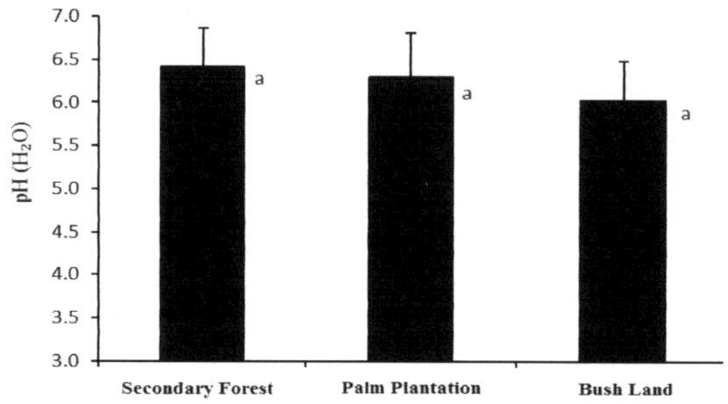

Figure 5. Comparison of pH (H_2O) on peatland vegetation of bush land (n = 10), palm plantation (n = 10), and secondary forest (n = 10). Sign different letters (a, b, c) indicate significant differences at α = 0.05.

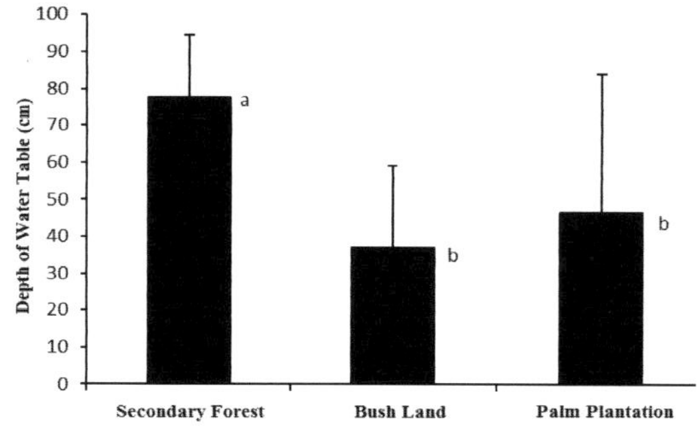

Figure 6. Comparison of the depth of the ground water level in the peat land with vegetation such as shrubs (n = 10), palm plantations (n = 10), and secondary forests (n = 10). Sign different letters (a, b, c) indicate significant differences at α = 0.05.

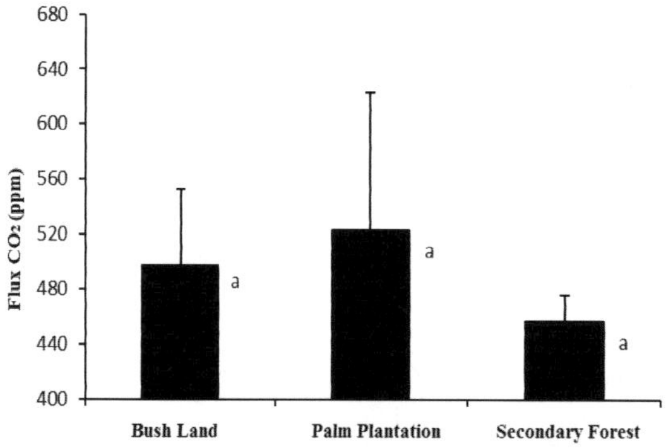

Figure 7. Average of CO_2 fluxes of bush land (n = 9), palm plantations (n = 9), and secondary forest (n = 9). Same letter shows no significant differences at α = 0.05.

Effect of the peat soil temperature toward CO_2 fluxes of measured for nine time replicates were done with different methods of analyses. Fig 8 and Fig. 9 showed the measurement of CO_2 flux for the bush land and secondary forest using Infrared Gas Analyzer. While, the CO_2 Flux measurement of palm plantations showed in Fig. 10.

3.2. Effect of soil temperatures

Table 2. Measurements of CO_2 Flux and depth of water table on the bush land using *Infrared Gas Analyzer*.

Distance between sample and canal (m)	Depth of water table (cm)*	CO_2 Flux (μmol m^{-2} sec^{-1})*
5	77,3	7,5
25	70,6	5,2
40	64,2	6,4

*) 17-20 replicates.

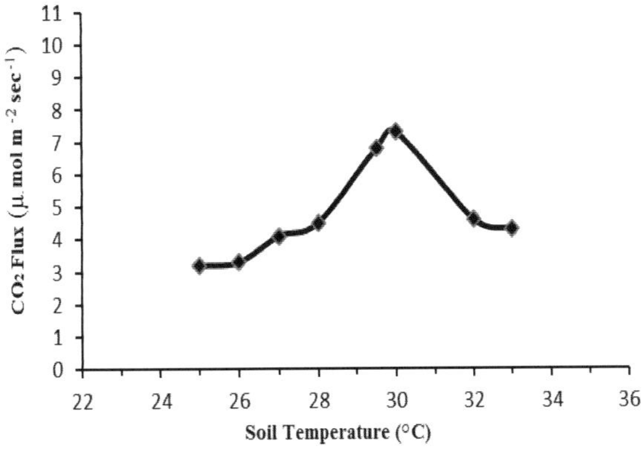

Figure 8. Effect of soil temperature toward CO_2 flux conducted in bush land use Infrared Gas Analyzer.

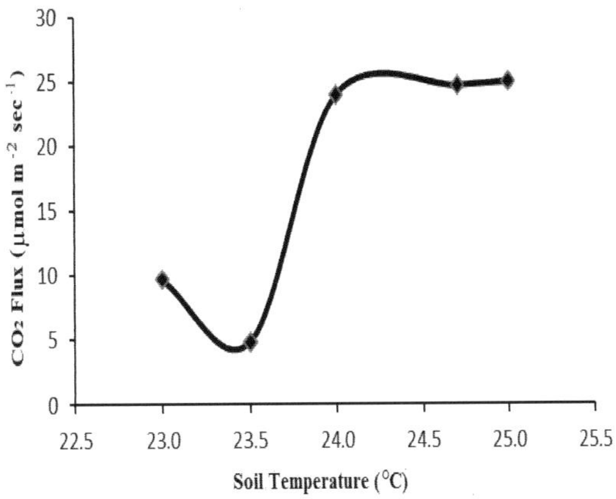

Figure 9. Effect of soil temperature toward CO_2 flux of secondary forests using Infrared Gas Analyzer.

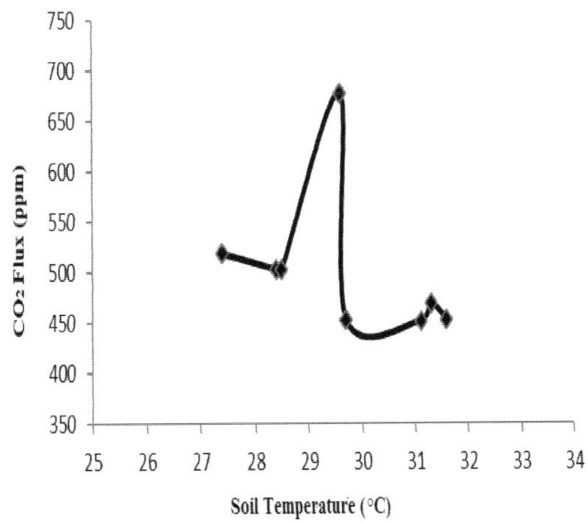

Figure 10. Effect of soil temperature toward CO_2 flux carried on palm plantations using Gas Chromatography.

3.3. Discussions

Based on the degree of decomposition of peat soil, the general types of peat in the study area belong to the Haplosaprist [9]. At the same location, peat soil in the study area has a dark brown color to dark black [10]. When viewed from the level of fertility of soil, peat soil is classified into types oligotrophic with the fertility is low and reacted sourly until very surly.

The average depth of the water table for bush land is 37.0 cm, 46.7 cm palm plantation and secondary forests of 77.7 cm. The depth of water table of the land affected by the type of cover vegetation growing on it. Secondary forest is a type of land that has deeper water. This is due to the characteristics of the soil on that land have pores larger than the bush land and palm peat structure. In addition, the number of roots of vegetation also affect the availability of water adsorbed on the land.

Peat water has a characteristic brown and sour with a pH range of about 3-5 [11]. The acidity of peat water caused by organic acids content in peat colloids. According [8] increasing the pH of peat caused by the length of the land use, application of fertilizers, as well as the high level of groundwater. The average pH of the water of bush land, palm plantations and secondary forests were 6.3, 6.0, and 6.4, respectively. Results of research conducted by [2] on the same type of peat (oligotrophic) in Central Kalimantan have the higher pH values ranging from 3.27 to 3.75. The low level of acidity of any land in this study because the measurements were conducted during the rainy season so that the pH of the water is no longer pure pH of the mixture of peat water but rainwater to seep into the peat.

Based on the results of temperature, bush land is the area that have a higher soil temperature of 31.1°C, when it compared to palm plantations and secondary forests i.e. 29.7°C and 24.2°C. The high temperature of the soil in bush land caused by the vegetation cover of the land such as ferns, grasses, as well as various types of broadleaf plants. The results of measurements of soil temperature at this research is not much different from the research carried out by [12] on a bush and secondary forest with soil temperature around 34.5°C and 28.1°C.

3.4. Emissions of greenhouse gases (GHG)

The big difference from the type of vegetation that covered the floor of peat and the depth of the ground water level, soil temperature, and pH (H_2O) creates large flux generated by each land supposed to be different. Otherwise, in this study, the difference in the flux of each land is not too sharp. Duncan test results conducted on $\alpha = 0.05$ also shows that there is no significant difference between the land. CO_2 flux measured from bush land, palm plantations, and secondary forest were 497.4 ppm, 523.2 ppm; and 457.2 ppm, respectively. CO_2 flux generated in this study rated exceed the concentration limit of clean air. As reported by [13] that the boundary between the clean air has a range of 310-330 ppm, while in this area in a range of 350-700 ppm. It means that the air around this area polluted.

Generally, peatlands as secondary forest produce CO_2 fluxes higher than the bush land and palm plantations. This condition occurred because there were a lot of peat sorb in the process of weathering. Furthermore, it is also because of the respiration of the roots of the forest vegetation. The same result was reported by [14] on research conducted in Central Kalimantan. In addition, the differences occur in this study was also supported by other factors such as time and method used for measuring gas samples at different secondary forest with methods that do to the bush and palm plantation. The occurrence of the time difference and the current method of sampling due to the limited number of the chamber at the time so that the sampling time in the bush land and palm plantation early do than secondary forests.

The measurements of CO_2 flux at each peatland showed a positive correlation between soil CO_2 flux with temperature, pH (H_2O), and the depth of the ground water level. An average depth of the water level were 56.0 cm, 31.7 cm and 29.8 cm, respectively. As stated [14], the higher the production of CO_2 emitted flux peat caused the changes of soil conditions from anaerobic into aerobic. It causes change in the rate of organic matter decomposition process. Decomposition during aerobic conditions flux of CO_2 produced 50 times faster than anaerobic. Aerobic conditions will trigger the availability of oxygen in the soil so that the microbial decomposition as media gets more energy to accelerate the process of mineralization of organic C to produce CO_2 [15]. In contrast to anaerobic conditions, there is no oxygen availability due to water-saturated soil conditions so that the process of mineralization of organic C tends to produce CH_4 than CO_2.

Temperature of peat soil in bush land ranging between 25-33°C and palm plantations from 27.5 to 31.5°C. The results of both the land of each field have different optimum temperature limits. At the optimum temperature, the enzyme system works very well and remain stable for a longer time. Otherwise, at low temperature enzymes also stable but they cannot serve as biocatalyst. While at higher temperatures, enzymes undergo denaturation that causes loss of the catalytic properties (Salampak et al., 2014). The secondary forest covered lot of trees, so the influence of temperature changes on the release of CO_2 flux not seen.

4. Conclusions

Based on the research, it can conclude as follows:

1. The average flux of CO_2 emitted by palm plantations was 523.2 ppm, higher than the bush land and secondary forests of 497.4 ppm and 457.2 ppm, respectively. It does not show any significant differences at $\alpha = 0.05$.
2. The pattern of the relationship between CO_2 flux with depth of the ground water level, pH (H_2O), and the temperature of the ground at each land has a positive correlation, except for the relationship between of ground water depth with CO_2 emissions on secondary forests.

5. Acknowledgement

Funding for this project was provided by Center for Southeast Asian Studies Kyoto University, Japan. It is also supported by University of Riau's Rector and Faculty of Maths and Natural Sciences Dean.

6. References

[1] Wahyunto, S. Ritung, Suparto, and Subagjo H 2003 *Sebaran Gambut dan Kandungan Karbon di Sumatera dan Kalimantan*. Wetlands International-Indonesia Programme dan Wildlife Habitat Canada, Bogor.

[2] Mubekti 2011 Studi pewilayahan dalam rangka pengelolaan lahan gambut berkelanjutan di Provinsi Riau. *Jurnal Sains dan Teknologi Indonesia* **13**(2): 88-94.

[3] Purnasari T, Muhammad A, and Salbiah D 2012 Keanekaragaman dan Biomassa Rayap Tanah di Kebun Kelapa Sawit dan Kebun Pekarangan Pada Lahan Gambut di Kawasan Bukit Batu, Riau. Fakultas Matematika dan Ilmu Pengetahuan Alam Universitas Riau, Pekanbaru, 2012.

[4] Jauhiainen J, Takahashi H, Heikkinen J E P, Martikainen P J and Vasander H 2005 Carbon Fluxes from a Tropical Peat Swamp Forest Floor. *Global Change Biology* **11**: 1788–1797.

[5] Yulianti N 2009 Cadangan karbon lahan gambut dari agroekosistem kelapa sawit PTPN IV Ajamu, Kabupaten Lahan Batu, Sumatera Utara. *Tesis*. FP Institut Pertanian Bogor, Bogor, 2009.

[6] Jabro J D, Sainju U, Steven W B, and Evans R G 2008. Carbon dioxide flux as affected by tillage and irrogation in soil converted from perennial forages to annual crops. *JOURNAL OF Environmental Management*. **88**: 1478-1484.

[7] Rumbang N, Radjagukguk B, and Prajitno D 2009 Emisi karbon dioksida (CO_2) dari beberapa tipe penggunaan lahan gambut di Kalimantan. *Jurnal Ilmu Tanah dan Lingkungan* **9**(2): 95-102.

[8] Dariah A, Jubaedah, Wahyunto, and Pitono J 2013 Pengaruh Tinggi Muka Air Saluran Drainase, Pupuk, dan Amelioran Terhadap Emisi CO_2 Pada Perkebunan Kelapa Sawit di Lahan Gambut. Balai Penelitian Tanah, Bogor.

[9] Soil Survey Staff. 1998. *Keys to Soil Taxonomy. 8th Edition*. Agency for International Development United States Departement of Agriculture Soil Management Support Services. Virginia Polytechnic Institute and State University. 716 p.

[10] Nasrul B. 2010 Penyebaran dan Potensi Lahan Gambut di Kabupaten Bengkalis untuk Pengembangan Pertanian. Fakultas Pertanian Universitas Riau, Pekanbaru.

[11] Yusnimar, A. Yelmida A, Yenie E, Edward H S and Drastinawati 2010 *Pengolahan Air Gambut dengan Bentonit*. Fakultas Teknik Kimia Universitas Riau, Pekanbaru.

[12] Tono S. Wawan, and Amri I A 2011 Fluks CO_2 pada Berbagai Kondisi Hutan Rawa Gambut di Areal Konsesi PT. Diamond Raya Timber Kecamatan Bangko Kabupaten Roakn Hilir. Fakultas Pertanian, Universitas Riau, Pekanbaru.

[13] Huboyo S H, Budihardjo. A M 2008 *Pencemaran Udara*. Fakultas Teknik Universitas Diponegoro, Semarang.

[14] Salampak, Sustiyah, and Amelia V 2014 *Fluks Gas Karbon Dioksida Pada Tanah Gambut Pedalaman di Kalampangan, Kalimantan Tengah*. Fakultas Pertanian, Universitas Palangkaraya, Kalimantan Tengah.

[15] Handayani E 2009 Emisi Karbon Dioksida (CO2) dan Metan (CH4) Pada Perkebunan Kelapa Sawit di Lahan Gambut yang Memiliki Keragaman dalam Ketebalan Gambut dan Umur Tanaman. *Thesis*. FMIPA Institut Pertanian Bogor, Bogor.

IOP Conf. Series: Earth and Environmental Science **68** (2017) 012003 doi:10.1088/1755-1315/68/1/012003

A Mini-review of Carbonaceous Nanomaterials for Removal of Contaminants from Wastewater

S J Shan[1,2], Y Zhao [1,2,*], H Tang [1,2] and F Y Cui[1,2]

[1] State Key Laboratory of Urban Water Resource and Environment, Harbin 150090, China
[2] School of Municipal and Environmental Engineering, Harbin Institute of Technology, Harbin 150090, China

Abstract: Recently, carbonaceous nanomaterials have been extensively studied for all sorts of contaminants removal from wastewater due to its extraordinary and tunable properties. The present and potential applications of these nanomaterials in wastewater treatment include adsorption, photocatalysis, disinfection, membranes process and other utilizations such as monitoring and desalination. In this paper, we concisely overview the current advances in carbonaceous nanomaterials, covering the basic information upon these nanomaterials (e.g., categories, structure, versatile properties) and their roles for removing diverse contaminants in different applications. Furthermore, the challenges and prospects for further utilizations are also outlined.

1. Introduction

With the rapid development of urbanization and industrialization, numerous contaminants were discharged from industrial production and households into aquatic environment. These contaminants include nutrients, toxic substances such as heavy metals, organic compounds, dyes, parasites and many other complex compounds, which have caused severe influence both on the health of ecological environment and human beings. Consequently, removal of toxic pollutants from wastewater is urgently needed. For this purpose, a number of conventional methods like precipitation, coagulation and other physicochemical technologies have been commonly employed. However, these approaches have been restricted by many factors, such as the increasing stringent water quality standards, the appearance of emerging contaminants, lower removal efficiency, complicated operating conditions, and expensive costs. For instance, chlorination could generate various disinfection by-products (DBPs) with a high mutagenic and carcinogenic risk as well as undesirable odors and tastes during the treatment process. Ion exchange, require complex condition actions and exhibit poor universality, which cannot meet the demands of rigorous water quality standards.

Recently, advance in nanoscience and nanotechnology have suggested that many environmental concerns involving wastewater decontamination could be solved or greatly diminished through the utilization of several promising nanomaterials. A variety of nanomaterials with unique functionalities such as nanoadsorbents, nanocatalysts, nanostructured membranes, have been considered as efficient, economical, and environmental-friendly substitutes to the current wastewater treatment agents.

Here, we provide a succinct review of recent advances in carbonaceous nanomaterials during waster decontamination. The utilizations of carbonaceous nanomaterials are strictly discussed on the basis of their functions in unit process and combined with other treatment approaches. The simple overview concerned the synthesis routes and brilliant properties of these exceptional nanomaterials are presented.

Furthermore, the gaps which limited their full-scale applications and the researches need to tackle it are also considered. The environmental fate along with other potential risks of these nanomaterials are beyond the scope of this article and thus would not be discussed at great length.

2. Carbonaceous nanomaterials

Carbonaceous nanomaterials integrate the distinctive properties of sp^2 hybridized carbon bonds with extraordinary physical-chemical properties at the nanoscale, generally exhibit excellent chemical, mechanical and electrical properties. They have attracted wide attentions in many applications, including environmental remediation, contaminants transformation, drug delivery, supercapacitors, catalysis, etc. With the booming developments of nanotechnologies, roles of carbonaceous nanomaterials would be more and more significant in future life. Nanoporous activated carbon (NAC) is quite common and will not describe in this section.

2.1. Fullerenes

Fullerenes are regarded as the third allotropic form of carbon materials since the discovery in 1985. They are enclosed cage-like structures comprising of 12 pentagonal rings and unspecific number(other than one) of 6-members rings, generally appear in the shape of 6-members rings in which carbon atoms placed. For example, C_{60} is composed of 32 faces with 12 pentagons and 20 hexagons, all carbon atoms are arranged at the 60 vertices of a truncated icosahedron, is highly symmetric[1]. Structures with fewer hexagons exhibit greater sp^3 bonding character, higher strain energies, and availability of more reactive carbon sites. Moreover, the chemical properties of fullerenes is abundant and various, which allow modifications of them for exploiting relevant environmental applications. Several measures such as covalent, supramolecular and Endohedral transformations have been taken to achieve the goal for practical use. Approaches that introducing -OH and -COOH groups could enhance the dispersion of pristine fullerenes, combination with polymers would alter its electrical and mechanical properties.

2.2. Graphene-based nanomaterials

2.2.1. Graphene. Graphene is an infinite two-dimensional (2D) monolayer consisting of sp^2-hybridized carbon atoms. It could be viewed as the fundamental building block for other carbon allotropes such as fullerenes and carbon nanotubes. It also exhibit intrinsic corrugations and some defects such as topological defects, vacancies, which would influence the reactivity of graphene with others. And the theoretical surface area of graphene could reach to 2630 m^2/g, is much higher than many materials. With these special structural properties, graphene has shown standout attributes and were studied in many applications. Various approaches were employed to synthesize graphene nanosheets and can be broadly categorized two sorts, "top-down" approach and "bottom up" approach, which include exfoliation of natural graphite and reduction from GO[2].

2.2.2. Graphene oxide (GO) and reduced graphene oxide (RGO). GO, can be made from chemical exfoliation of graphite by reactions, is a two-dimensional sheet with abundant oxygenated functional groups on their basal plane and edges. The fabrication process of GO breaks the π-conjugated network and thus make it highly water-dispersible[3], meanwhile, reduce the aggregation among individual GO sheets, display a wider potential in wastewater treatment. And GO itself has attracted significant attentions due to many intriguing properties and its role as a promising precursor for bulk production of graphene-based nanomaterials. Synthesis approaches of GO contains the Brodie, Staudenmaier, Hummers method, or modified methods based on them [2].

The structure of RGO is between graphene and GO, there are only a few functional groups embedding on the RGO surface, could be reduced from GO by solvothermal and hydrothermal methods [6]. The enhanced electronic and conductive properties make it a prospective candidate for pollutants removal in wastewater treatment process.

2.2.3. Graphene nanocomposites. In recent years, abundant polymers and metal oxides were incorporated to the graphene-based sheets, developed as nanocomposites. These nanocomposites exhibit standout structural performance and multifunctional properties by combing both components' characteristics. And it should be noted that integrated nanocomposites are not merely the sum of different components, but instead a new nanomaterials with entirely new properties [2]. The functionalization process could facilitate the dispersion and then prevent the aggregation of graphene sheet in the aqueous phase, and endow them with some superior chemical and physical attributes, which may contribute to many practical utilizations. For example, magnetic nanocomposites of graphene could be easily separated from treated water and thus reduce serious recontaminations, but it is difficult for separation of original graphene and GO. The incorporated nanocomposites could be more selective and bind the desired pollutants more tightly based on its structures.

2.3. Carbon nanotubes (CNTs)

CNTs, also called buckytubes, are cylindrical carbon molecules with unique properties. They can be divided into single-walled CNTs (SWCNTs) and multi-walled CNTs (MWCNTs) on the basis of the tubes number in the CNTs, where SWCNTs are made of single layer carbon atoms while MWCNTs comprise of multilateral graphene sheet that rolled upon itself with dozens of concentric tubes [4]. As with graphene-based nanomaterials, CNTs exhibit unique properties such as large specific area, rich hollows and layered structure, which bring about a wide variety of remarkable applications in wastewater treatment, including the removal of organic contaminants and heavy metals. There are various methods could be employed to synthesize CNTs, for instance, chemical vapor deposition (CVD), the arc-discharge method, flame synthesis method and other modified methods[4]. Other nanocomposites incorporated with polymers and metal oxides could be prepared by hydrothermal method, microwave irradiation and other approaches.

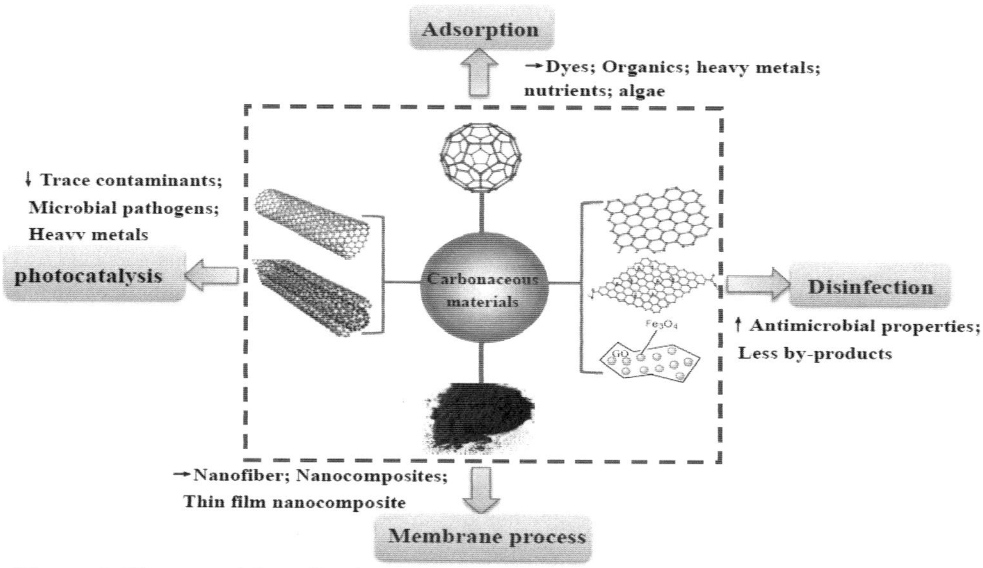

Figure 1. The potential applications of carbonaceous nanomaterials in wastewater treatment. Carbonaceous nanomaterial could be divided into four categories: fullerenes, graphene-based nanomaterials, CNTs and NAC. All these nanomaterials exhibited great potential in many aspects of wastewater treatment such as adsorption, photocatalysis, disinfection and membrane process, etc.

3. Applications of carbonaceous nanomaterials in wastewater treatment

Carbonaceous nanomaterials showed great potential in wastewater treatment (see **Figure 1.**), would be discussed concisely in this part.

3.1. Adsorption

Adsorption has been considered as the most promising method for contaminants removal from aqueous phase due to its low cost, simple operation and no by-products. Nevertheless, the efficiency of conventional adsorbents is always limited and thus inhibit the pollutants removal process. Compared to traditional adsorbents, nanostructured materials possess a higher specific area and associated sorption sites along with tunable surface characteristics, which give them some strengths such as higher efficiencies and rapid adsorption rates over a broad pH range.

3.1.1. Removal of organic contaminants. Graphene-based nanomaterials have displayed exceptional adsorption capacity toward various organic compounds, covering dyes[5], antibiotics[6], polycyclic aromatic hydrocarbons (PAHs)[7], phenolic compounds[8], pesticides[9], and so forth. In aqueous phase, adsorption capacities of chlorpyrifos, endosulfan, malathion on RGO could reach to 1200, 1100, 800mg/g, respectively [9]. Other carbonaceous nanomaterials such as CNTs, fullerenes and nanocomposites are also effective to remove organics from wastewater and have been widely studied and extensively reviewed. Their strong adsorption capacity are mainly attributed to the incredibly specific surface area, abundant pore size distribution, and feasible surface properties, which make them interact vigorously with organic compounds, via hydrophobic effects and EDA interactions, hydrogen bonding interactions, etc.
Additionally, studies pointed out that environmental conditions and surface characteristics of different type adsorbents would affect the adsorption capability, and thus a little departures could be found among adsorption performance. The aggregation of CNTs in aqueous phase create a lot of interstitial spaces and grooves, which belong to high-energy adsorption sites and could contribute to organic contaminants removal. For GO, enormous oxygen-containing functional groups (e.g.,–COOH and–OH) on the edge would weaken the hydrophobic effect with nonpolar organics, and suppress the adsorption on it. For NAC, large-sized molecules (e.g., algal toxins) are inaccessible due to its micro-porous structures, but can be benefit for the relatively small molecules.

3.1.2. Removal of heavy metals. Many researchers have launched extensive studies concerning the adsorption characteristics of heavy metals onto GO and oxidized CNTs, indicating that both GO and oxidized CNTs showed great adsorption performances and rapid adsorption rates for metals removal[10, 11]. The ability of carbonaceous nanomaterials to adsorb different heavy metals (e.g. Cd^{2+}, Pb^{2+}, Zn^{2+}, Cu^{2+}, Co^{2+}) have been reviewed critically[12, 13]. The maximum adsorption capacities of Cd^{2+} and Co^{2+} on GO are 106.3, 68.2mg/g, respectively, which higher than other nanoadsorbents such as NAC[13]. The electrostatic attraction and chemical bonding such as ion exchange dominated the adsorption process where the surface functional groups of GO and oxidized CNTs provide the major adsorption sites to heavy metals. Fast adsorption kinetics could be attributed to the highly accessible adsorption sites and the short intraparticle diffusion distance. Other nanomaterials and nanocomposites also have attracted increasing attentions for heavy metals removal.

3.2. Photocatalysis
Photocatalytic technique is an advanced oxidation process for removal of trace contaminants and pathogens from wastewater, which have been considered as an useful pretreatment approaches for toxic substance and non-biodegradable pollutants to enhance its biodegradability [14]. Currently, many researches focused the photocatalysis enhancement by carbonaceous nanomaterials have been rigorously examined. One hot topic is the utilization of carbonaceous nanomaterials (e.g., CNTs, fullerenes, RGO) in synergy with catalyst particles (TiO_2, etc.). By incorporating TiO_2 onto individual graphene nanosheets with even distribution could gain desirable properties from their fabrication process, reinforce the photocatalysis performance for target contaminants. Yang et al synthesized a series of TiO_2-graphene, -CNTs, -fullerenes nanocomposites photocatalysts to oxidize benzyl alcohol to benzaldehyde, indicating that integrated nanocomposite could broaden the adsorption edge to the visible light region. Apart from TiO_2/carbonaceous nanocomposites, several other hybrid photocatalysts have been investigated for the detoxification of organic contaminants, such as ZnS-RGO [15], ZnO/graphene, etc. Indeed, CNTs and graphene would not only provide high-surface area support and immobilization for photocatalysts particles (e.g., TiO_2). And the presence of

carbonaceous nanomaterials could enhance the photocatalysis performance through three primary mechanisms: (1) providing high quality adsorption active sites; (2) improving the migration efficiency of photo-induced electrons and delaying electron-hole recombination; (3) tuning the band gap or photosensitization.

3.3. Disinfections

Disinfection is an effective method to remove organic contaminants and biological pollutants such as pathogens from wastewater sources. While conventional technologies (e.g., chlorination, ozonation) could form hazardous DBPs in the decontamination process and pose a threat to human health. Advance in nanotechnology provide a new perspective to overcome weakness of these old methods.

Carbonaceous nanomaterials exhibit strong antibacterial activity and lower oxidation ability, could effectively inactivate pathogens under visible-light irradiation or direct contact, have low tendency to form DBPs. Graphene-based nanocomposites were widely investigated as disinfectants in the last few years. The magnetic graphene developed by Ganesh et al[16] exhibited disinfection activity (40ug/L) toward *E.col* with 100% killing efficiency as well as desirable low toxicity toward zebrafish. The magnetic RGO functionalized with glutaraldehyde nanocomposite (MRGOGA) synthesized by Wu et al.[17] also showed the brilliant disinfection ability towards *S. aureus* and *E.col*, up to 99% of gram-positive and gram-negative bacteria were killed effectively in 10 minutes upon near-infrared laser irradiation. The antibacterial activity, large specific surface area and strong conductivity of graphene-based nanomaterials enabled their use in disinfection applications. And the antibacterial mechanism of these carbonaceous nanomaterials was reported to involve two aspects: (1) destroying the integrity of cell membranes upon direct contact; (2) disturbing the particular microbial progress via oxidative stress. Furthermore, additional researches associated with maximizing disinfection performance and environmental effects of nanomaterials should be paid more attention.

3.4. Membrane process

Membrane technique is considered an effective and economical approach in many fields of wastewater treatment, its performance significantly depends on the selectivity and permeability of the membrane materials. High energy consumption caused by membrane fouling and other problems limited the wide application of this technique. Considerable studies have reported that incorporation of carbonaceous nanomaterials into membranes make it possible to improve the selectivity, permeability and fouling resistance of membranes.

CNTs could be incorporated with other polymers or nanomaterials to develop multifunctional membranes. Song et al. [18] infused CNTs into a double-skinned thin film nanocomposite (TFN) membranes and characterized the morphology, structure, permeation performance and antifouling ability of the integrated membranes. It was found that TFN membranes exhibited higher water flux and remarkable antifouling capacity by incorporating CNTs. This result was consistent with research of Ho et al. [19], incorporation of GO and oxidized CNTs have significant effect on membrane characterization including surface charge, porosity, pore size structure, mechanical strength, toughness and other indicators. Hence, the mixed-matrix membranes increased the membrane hydrophilicity and antifouling properties, meanwhile, improved the rejection rate of both organic pollutants and inorganic matters. Free-standing carbonaceous nanofiber membranes are the another type of carbonaceous filters, which possess huge specific surface areas, high degree of uniformity and massive active sites. They combined the excellent performance of carbonaceous nanofiber and the advantage of membranes blocking with strong permeability and antifouling capacity, beyond the traditional adsorption technique. Carbonaceous nanofiber membranes could remove dyes (MB) quickly at a very high flux of 1580 $L \cdot m^{-2} \cdot h^{-1}$, which is much higher than many commercial ultrafiltration membranes with similar behaviors [20]. Besides, other catalysts also could be doped into carbonaceous membranes to remove contaminants through various mechanisms simultaneously.

3.5. Other applications

Apart from the above-mentioned applications, there are still several utilizations of carbonaceous nanomaterials in wastewater treatment, which include sensing, monitoring and desalination. Nurul et al. [21] incorporated MWCNTs and single layer graphene (SLG) nanoparticles into cellulose triacetate (CTA) polymer matrix and developed a MWCNTs-MMM and SLG-MMM nanocomposites, respectively. It illustrated that both nanomaterials were successfully applied for PAHs detection in wastewater samples with a detection limit of 0.02-0.09 ng/mL, were comparable or even better than standard SPE technique. The adsorption behavior that occurred on the active sites of nanomaterials' surface took charge of the microextraction process. Researches involving detection of other trace contaminants and pathogens along with other potential utilizations are growing rapidly, and we believe that more comprehensive and full-scale applications of carbonaceous nanomaterials would be found in the future treatment process.

4. Challenges and prospects for further application

Carbonaceous nanomaterials possess high specific surface areas, outstanding electrical, optical, thermal and chemical activity; have been considered one of the most prospective candidates to remove chemical and biological contaminants from wastewater. Previous studies have already concerned the potential application of these nanomaterials for wastewater treatment in lab scale. However, the future developments and full-scale applications of these nanomaterials still face a wide variety of challenges and more thorough studies are certainly needed.

1) The current synthesis methods for carbonaceous nanomaterials (e.g., RGO) is still complicated and low-efficient, although numerous studies have been conducted to tackle it. More simple, robust and efficient fabrication methods are urgently needed. Meanwhile, commercial large-scale production of carbonaceous nanomaterials is challenging and need to be addressed for broad range applications.

2) The agglomeration of CNTs and graphene-based nanomaterials in aqueous phase is another weakness in water decontamination. Aggregated nanomaterials would decline the surface area as well as active sites and hence affect the accessibility to them, resulting in decreased efficiency on pollutants removal. Nanomaterials modified with various functional groups and metal oxides have been employed to overcome them, and further researches should focus more on the targeted modification and enhance the removal efficiency as well as selectivity and affinity toward specific contaminants.

3) Almost all the studies were conducted under laboratorial conditions, the performances of carbonaceous nanomaterials in treating real natural and wastewater from industries need to be tested. Further researches should be launched under more realistic conditions to evaluate the efficiency and applicability of diverse carbonaceous nanomaterials in wastewater treatment. Besides, the long-term performance of carbonaceous nanomaterials should also be taken into consideration.

4) The cytotoxicity of carbonaceous nanomaterials for human beings, living things and ecosystem have not been investigated intensively, the gap between potential environmental effects and carbonaceous nanomaterials' properties should be filled as soon as possible. In order to guarantee the environmental and public health, risk assessment for carbonaceous nanomaterials is urgently needed with the increasing emission of these nanomaterials.

Carbonaceous nanomaterials have many advantages as well as limitations in wastewater treatment, it is indeed potential nanomaterials for solving diverse environmental problems.

5. Conclusions

Water decontamination is one of the most essential tasks for both environmental protection and wastewater reuse. Among various wastewater treatment techniques, using carbonaceous nanomaterials to remove diverse pollutants have attracted wide attentions in many sectors and momentum. The unique properties of carbonaceous nanomaterials bring about great chance to revolutionize wastewater treatment. Although most of highlighted applications of carbonaceous nanomaterials are still in the laboratory stage, some have proved marvelous properties in the real wastewater treatment process. Further work concentrated on the modification of these nanomaterials is a helpful approach to accelerate the potential applications by improving intrinsic properties and economic benefits of carbonaceous nanomaterials, display a brighter future in wastewater treatment.

6. Acknowledgement

This work was supported by National Natural Science Foundation of China (Grant No. 51278147 and 50808052).

7. Reference

[1] Anafcheh M, Ektefa F 2015 Cyclosulfurization of C_{60} and C_{70} fullerenes: a DFT study *Struct. Chem.* **26** 1115-24
[2] Chowdhury S and Balasubramanian R 2014 Recent advances in the use of graphene-family nanoadsorbents for removal of toxic pollutants from wastewater *Adv Colloid Interface Sci* 35-56
[3] Kim J, Cote L J, Huang J 2012 Two dimensional soft material: new faces of graphene oxide *Acc. Chem. Res.* **45** 1356-64
[4] Ibrahim K S 2013 Carbon nanotubes properties and applications: a review. *Carbon lett* **14** 131-44
[5] Ai L, Zhang C and Chen Z 2011 Removal of methylene blue from aqueous solution by a solvothermal-synthesized graphene/magnetite composite *J Hazard Mater.* **192** 1515-24
[6] Gao Y, Li Y, Zhang L, Huang H, Hu J and Shah SM 2012 Adsorption and removal of tetracycline antibiotics from aqueous solution by graphene oxide *J Colloid Interface Sci.* **368** 540-6
[7] Wang J, Chen Z and Chen B 2014 Adsorption of polycyclic aromatic hydrocarbons by graphene and graphene oxide nanosheets *Environ Sci Technol.* **48** 4817-25
[8] Xu J, Wang L and Zhu Y 2012 Decontamination of bisphenol A from aqueous solution by graphene adsorption *Langmuir* **28** 8418-25
[9] Maliyekkal S M, Sreeprasad T S, Krishnan D, Kouser S, Mishra A K and Waghmare U V 2013Graphene: a reusable substrate for unprecedented adsorption of pesticides *Small* **9** 273.
[10] Sitko R, Turek E, Zawisza B, Malicka E, Talik E, Heimann J, et al. 2013 Adsorption of divalent metal ions from aqueous solutions using graphene oxide *Dalton Transactions.* **42** 5682-9
[11] Tofighy M A and Mohammadi T 2011 Adsorption of divalent heavy metal ions from water using carbon nanotube sheets *J Hazard Mater.* **185** 140-7
[12] Mubarak N M, Sahu J N, Abdullah E C and Jayakumar N S 2013 Removal of heavy metals from wastewater using carbon nanotubes *Sep. Purif. Rev.* **43** 311-38
[13] Zhao G, Li J, Ren X, Chen C and Wang X 2011 Few-layered graphene oxide nanosheets as superior sorbents for heavy metal pollution management *Environ Sci Technol.* **45** 10454-62
[14] Qu X, Alvarez P J and Li Q 2013 Applications of nanotechnology in water and wastewater treatment *Water Res.* **47** 3931-46
[15] Thangavel S, Krishnamoorthy K, Kim S J and Venugopal G 2016 Designing ZnS decorated reduced graphene-oxide nanohybrid via microwave route and their application in photocatalysis *J. Allo. Com.* **683** 456-62
[16] Gollavelli G, Chang C C and Ling Y C 2013 Facile synthesis of smart magnetic graphene for safe drinking water: heavy metal removal and disinfection control *ACS Sustainable Chem. Eng.* **1**
[17] Wu M C, Deokar A R, Liao J H, Shih P Y and Ling Y C 2013 Graphene-Based photothermal agent for rapid and effective killing of bacteria *ACS Nano.* **7** 1281-90
[18] Song X, Wang L, Tang C Y, Wang Z and Gao C 2015 Fabrication of carbon nanotubes incorporated double-skinned thin film nanocomposite membranes for enhanced separation performance and antifouling capability in forward osmosis process *Desalination.* **369** 1-9
[19] Ho K C, Teow Y H, Ang W L and Mohammad A W 2017 Novel GO/OMWCNTs mixed-matrix membrane with enhanced antifouling property for palm oil mill effluent treatment *Sep. Purif. Technol.* **177** 337-49
[20] Liang H W, Cao X, Zhang W J, Lin H T, Zhou F, Chen L F, et al. 2011 Robust and highly efficient free-standing carbonaceous nanofiber membranes for water purification *Adv. Funct. Mater.* **21** 3851-8
[21] Mukhtar N H and See H H 2016 Carbonaceous nanomaterials immobilised mixed matrix membrane microextraction for the determination of polycyclic aromatic hydrocarbons in sewage pond water samples *Anal Chim Acta.* **931** 57-63

IOP Conf. Series: Earth and Environmental Science **68** (2017) 012004 doi:10.1088/1755-1315/68/1/012004

Analysis of Indoor Air Pollution of Decoration and Control Measures

Li Yan

Energy and Electromechanical Engineering College, Hunan University of Humanities, Science and Technology, Lou Di, China

Abstract. Nowadays, the human health is closely related to quality of indoor air. This article analyzes the main types of pollution to indoor air and their harms to human health, and on this basis, it sets forth the prevention measures comprehensively and proposes advices to normalize industry standards.

1. Introduction

People now spend most of their time indoors, either in home or in office. As the modernization of life style, the time spent indoors is in majority, and even as high as 93%. According to WHO report, the death caused by outdoor air pollution is about 3.7 million each year; however the death caused by indoor pollution is 4.3 million. When comes to China, the average death caused by this is up to 11.1 million, which means about 304 per day, equal to the number of people who died due to traffic accidents every day. The indoor air pollution is also considered as third kind of air pollution in addition to pollution types of " fuliginous model "and" fuliginous model". Years ago, the United States Environmental Protection Agency has counted indoor air pollution as the 4th largest hazards to environmental health after air pollution, toxic chemicals from factory and water pollution; and it is also be considered by WHO, as one of top ten threats to health. In 1980s, the terms of SBS (Sick Building Syndrom), BRI (Building Related Illness) and MCS (Multiple Chemical Sensitivity) etc., have been arisen. In conclusion, the human health is closely related to indoor air.

2. Main sources of indoor air pollution

2.1. Chemical pollutants

This kind of pollutant comes from decoration materials of wood-based panel, carpet, paint, adhesive and furniture, etc.; these materials often contain toxic and hazardous substances with actual content that exceeding certain limits (tabulated as table 1).

2.2. Physical pollutants

This kind of pollutant comes from building ceramic, including tile, wash basin and water closet etc.; it also derived from the natural stones like granite, marble; and the electromagnetic radiation caused by electric appliances of radio broadcasting, television and so on.

2.3. Biological pollutants

The biological pollutant, which includes virus, bacteria, mildew, fungi, dust mites and organic substances etc., mainly comes from domestic waste, household appliances, modern office facilities, indoor plants, pets, indoor decoration and furniture. All of these contaminations could cause diseases like allergic asthma, anaphylaxis, influenza and pneumonia etc.

Table 1. Limits of Indoor Environmental Pollutants in Civil Construction Engineering

Contaminant	Class I Civil Construction Engineering	Class II Civil Construction Engineering
Radon/(Bq/m3)	200	400
Free formaldehyde/(mg/m³)	0.08	0.10
Benzene/(mg/m³)	0.09	0.09
Ammonia/(mg/m³)	0.2	0.2
TVOC/(mg/m³)	0.5	0.6

2.4. Outdoor pollution

There are 3 main factors as following, ①the industry pollutions of sulfur oxides, nitrogen oxides and so on;②the photochemical smog and dust etc.; ③the automobile exhaust containing pollution of nitrogen oxides and hydrocarbon etc.

2.5. Biological pollutants

The biological pollutant, which includes virus, bacteria, mildew, fungi, dust mites and organic substances etc., mainly comes from domestic waste, household appliances, modern office facilities, indoor plants, pets, indoor decoration and furniture. All of these contaminations could cause diseases like allergic asthma, anaphylaxis, influenza and pneumonia etc.

3. Main contaminants and hazards

3.1. Formaldehyde

Formaldehyde is a colorless, irritant and volatile gas, it mainly released from wood-based panels, furniture and wall materials. The Formaldehyde posses a significant danger to human health especially for eyes, nose and mouth; for long-term exposure through respiratory and skin, it cause intoxication with indications of headaches, dizziness and sick, and it also leads to dysfunction in lung, liver and immune system. The formaldehyde is carcinogenic with a release duration as long as 3~15 years [1] on basis of finding from Yokohama National University. In newly constructed houses, the indoor formaldehyde is always more than ten times of relevant standards, sometimes even more than over forty times. Some scholars compared standard value of formaldehyde concentration in GB/T 18880-2002 [2] and GB50325-2010 [3] to the maximum actual concentration of formaldehyde2.11[4]、1.22 [5]、1.45 [5] in house, office and public place, and found it was more than 2.1 times, 1.2times and 1.4times to reference limits. Some researches show that the content of newly renovated reading room are higher than newly renovated house, and the outdoor posses relative lowest concentration, however the furnish posses highest concentration of formaldehyde regardless of the types. The correlation between concentration and body response is tabulated as Table 2 [6].

3.2. Volatile organic compounds(VOC)

The primary sources of indoor VOC are interior decorations of paint, floor, wall paper and adhesive etc. According to reports from Environment Protection Agency, the indoor VOC in a stuffy room will leads to indoor pollution higher than 10 times of that of outdoor. When VOC reaches a certain concentration in room, it causes symptoms of headache, nausea, vomit and exhaustion etc.; and may even incurs serious consequences of tic, coma, inflicting damage to liver, kidney, brain and nervous system, causing hematological abnormality, memory decline, and leukemia etc.. So the VOC has also been called as 'invisible killer'. Many studies on pathogenic causes indicate pollution is serious in newly renovated house which leads to a higher incidence of leukemia in infants and toddlers [7]-[9].

Table 2. Formaldehyde on acute irritating reaction for human health in short time.

Body response	**Concentration of formaldehyde, mg/m³**

Slight asthma for children	0.06-0.07mg/m³
Off-odors and discomfort	0.1mg/m³
Irritation to eyes	0.5mg/m³
Throat ache	0.6 mg/m³
Nausea, vomiting, chest tightness, pulmonary edema	37.5 mg/m³
Death	60 mg/m³

3.3. Benzene compounds
The benzene compounds are mainly sourced from chemical construction materials of filler, paint and various organic solvents etc. This kind of contaminant often possesses characteristics of volatility, inflammability and explosivity of its vapors. It causes dizziness, headache, nausea and weakness under inhalation of high concentration of benzene, xylene within short time, and even leads to coma and death in severe cases. And for long-term exposure, it will result in chronic intoxication, insomnia, headache, mental sluggishness, failing memory, and may even leads to leukemia. Actually, WHO has recognized benzene compounds as carcinogenic material.

3.4. Radon
The Radon is colorless, tasteless and natural radioactive inert gas that mainly released by underground water, burning of fuel gas and construction materials of slag brick, cinder brick, granite, tile and cleaning facilities. The Rn can be closely conjugate fat to widely distribute in blood, nervous system and reticuloendothelial system, and then rise damage to cells and eventually leads to cancer. WHO recognize Radon as carcinogen, which could increase the probability of cancer, and may simultaneously cause damage to heath of next generation and even third generation.

3.5. Ammonia
Ammonia is a colorless gas and has a strong irritating odor. It generally sourced from materials of concrete anti-freezing, additives and whitener of furniture finishing, wood plate; ammonia is alkaline that will corrode and irritate the substance directly contacted, and after massive inhalation to human body, it will incurs symptoms of lacrimation, dizziness, cough, nausea, chest tightness and breathing difficulties etc., and even rise pulmonary edema in serious case.

4. Control measures to indoor air pollution

4.1. Prevention and sources controlling
The best way of controlling is to control the pollution headstream.

4.1.1. Green environmental protection design. Hold consciousness of environment-protecting by well considering of ventilation, lighting etc. indoor during construction, the materials should be employed rationally to save the quantity.

4.1.2. Green building and decoration materials. The green and natural materials are recommended to be employed for decoration; at same time the reliability of green materials should be ensured.

4.2. Initiative management to reduce dangers

4.2.1. Ventilation. It's better to keep windows open for ventilation and put off the moving into newly constructed house [8] so as to dilute the concentration of hazardous gas, generally to say, it is the most common, effective and economical way to prevention of indoor pollution. Actually there are three types of ventilation as natural one, mechanical and natural ventilation with auxiliary facilities; thereinto, the mechanical ventilation consuming energy, thus the installation of this kind equipment should be well considered at the designing of building so as to save energy.

4.2.2. Physical adsorption. The adsorption property of adsorbent could be employed to reduce the smell and concentration of hazardous gas effectively. The normal adsorbents include activated carbon, bamboo charcoal, silica gel, plaster and diatom and so on; thereof the activated carbon posses characteristics of good hydrophobicity, highly developed capillary structure, cheap and good adsorption properties, it is used frequently, what is more, it can be recycled after soaking in salty water and direct-sun exposure; although so many advantages, the weak absorption to inorganic gas still needs to be strengthened. Some researches show that the fibred activated carbon behaves high ability to absorb inorganic gas as well as organic gas, with absorption of more than 1~10 times[9] to normal activated carbon. And some researchers are now studying on some of new types of activated carbon, which possess absorption of 4 times to the Bamboo Activated Carbon and 1.5 times to Coconut Base Activated Charcoal [10].

4.2.3. Biological absorption. The green plants are widely applied for low price, easy operation and great varieties, the function of green plants are not only limited to absorbing pollutants, but also beautifies the environment. Some researches [11], [12] indicate the plants posses unique advantage in purifying air. For instances, the ivy could eliminate about 90% of benzene; the bracts of rhizoma arisaematis could eliminate 80% of benzene and 50% of trichloro ethylene; the Chlorophytum comosum could decontaminate about 96% of carbon monoxide, 86% of formaldehyde and nitrogen peroxide [13]; the zamioculcas zamiifolia rehd could absorb 88% of dimethyl benzene just within 72hours [14], [15], and it is also effect in absorption of benzene, toluene and ethylbenzene indoors [*16*]; the plants of Sansevieria trifasciata and Kalanchoe blossfeldiana could decontaminate 77% of toluene witin 72hours; all Chlorophytum comosum as well as Sansevieria hyacinthoides and Aglaonema commutatum could eliminate 70% of ethylbenzene [16]; in addition to these, the monstera, cactus and aloe also possess function of air purification.

4.2.4. Initiative management. Air depuration devices could also be used to absorb the indoor pollution gases. The devices could absorb, decompose and transform various hazardous gases, is workable for air purification. Furthermore, many scholars and scientists put forward some new methods such as photocatalysis with nano materials, plasma catalytic [9], photocatalyst, air catalytic technology, discharge-catalytic technology with nano plasma of Hydrogen peroxide, etc [17].

4.3. Strengthen supervision to ensure health
It is better to request an analysis report from the decoration company after decoration, and also employed professional organization to monitor the quality of indoor air before moving in.

5. Summary
Along with the continuous improvement of living standard, people are paying more and more attention to the quality of indoor air. From the source, we hope the national authority could establish suitable regulations to normalize the specification for construction and decoration materials, and prevent fake and low- quality product and the firms manufacture or marketing such products from approaching market. Meanwhile all of us should encourage manufacturers to support more and more green products, also the people shall improve awareness of health, green decoration, and adopt more effective approaches to decrease the danger of indoor air pollution.

6. Acknowledgments
Because of such short notice, there are certain shortcomings and mistakes that have not yet been found. Of course, my personal opinions still need further analysis and meticulous argumentation, and I hope experts and scholars can help me improve it and even correct them. I shall be very grateful.

7. References
[1] Jiang Jinqing 2005 Journal of Shaoxing University(Natural Science).**25** 42-45
[2] GB/T 18880-200 2002 *Specification of indoor air*（Beijing: China Standards Press）
[3] GB50325-2010 2010 *Control practices for indoor pollution in construction for civil use*（Beijing: China Quality Press）
[4] Huang Jidian 2012 *Journal of Environment and Health.* **29** 1049-50
[5] Liu Aiguo, Yan Jiang 2010 *J.Practical Preventive Medicine.* **17** 1318-19

IOP Conf. Series: Earth and Environmental Science **68** (2017) 012004 doi:10.1088/1755-1315/68/1/012004

[6] Li Weihua, Li lianquan, Wang junjie and Zhang nuan 2016 *J. Cleaning& Air conditioner technology.* **3** 71-75
[7] Sun Xiaodong, Chai Yihuan, Deng Haizhen etc. *J. Medical CASE* **5** 1-3
[8] Luo Daocheng, Yi Pingui, Cheng Anguo 2003 *J.China Safety Sci.* **3** 66-68
[9] LU Shu-zhi 2003 *J. Liaoning Chemical Industry.***32** 204-5
[10] Hu Liupin, Mo Kailin, Yang Lin etc. 2017 *J. Sichuan Forestry Sci.and Technol.* **28** 15,53-55
[11] Cao Hui 2008 *J. Northern Horticulture* **6** 150-55.
[12] Zhou Xiaojing, Liang Shuangyan, Jin Youju, etc. 2006 *J. Chinese Agricultural Science Bulletin.* **22** 229-31
[13] Wang Yinglin, Wu Gaodi 2015 *J. Real Estate Biweekby.* **16** 304
[14] Sriprapat W, Boraphech P, Thiravetyan P 2014 *J. Environ Sci. Pollut.* **21** 2603-10
[15] Sriprapat W, Thiravetyan P 2013 *J. Water,Air Soil Poll.***224** 1482-90
[16] Sriprapat W, Suksabye P, Areephak S, et al. 2014 *J. Ecotox Environfe.* **102** 147-51
[17] Zhang Cheng, Chen Chao 2016. *J. Analysis and Discussion.* **4** 242-44

Chemical Speciation and Health Risk Assessment of Fine Particulate Bound Trace Metals Emitted from Ota Industrial Estate, Nigeria

Winifred U Anake[1*], Godson R E E Ana[2], Akan B Williams[1], Omowunmi H. Fred-Ahmadu[1] and Nsikak U Benson[1*]

[1]Analytical and Environmental Chemistry, Unit, Department of Chemistry, Covenant University, Km 10 Idiroko Road, Ota, Ogun State, Nigeria.
[2]Department of Environmental Health Sciences, University of Ibadan, Ibadan, Nigeria.

Abstract. In this study carcinogenic and non-carcinogenic health risk due to exposure to $PM_{2.5}$-bound trace metals from an industrial area in Southwestern Nigeria was estimated. A four-step chemical sequential extraction procedure was employed for the chemical extraction of arsenic (As), cadmium (Cd), chromium (Cr) copper (Cu), manganese (Mn), nickel (Ni), and zinc (Zn). Samples were analyzed using inductively coupled plasma mass spectrometry (ICP-MS). Results reveal Cr and Cu as the most dominant exchangeable fraction metals, indicating possibility of their being readily soluble once $PM_{2.5}$ is inhaled. Cd and Cr record the highest bioavailability index of 0.7. The cumulative lifetime cancer risks due to inhalation exposure for adults (4.25×10^{-2}), children 1-6 years old (4.87×10^{-3}), and children 6-18 years old (1.46×10^{-2}) were found above Environmental Protection Agency's acceptable range of 1×10^{-6} to 1×10^{-4}. The hazard index values for all studied trace metals suggest significant potential for non-carcinogenic health risks to adults and children. The choice of chemical speciation as an essential tool in facilitating a better predictive insight on metal bioavailability and toxicity for immediate remediation action has been highlighted.

1. Introduction

In Nigeria, significant amount of atmospheric particulates have been produced in recent years due to industrialization, unregulated waste management, emissions from vehicles, biomass burning, and rapid urbanization [1], [2]. Enhanced levels of PM in industrialized communities have heightened pollution problems in such areas and have been reported [1]-[4]. This has led to continuous deterioration in local and regional air quality with considerable mortalities and economic loss in these communities. Ota industrial estate (latitude 6° 32'N, longitude 2° 57'E) was established in 1983 as one of largest industrial hub in Ota, Nigeria. Over the past years, about thirty industries have been established and are domiciled within this industrial estate. The industrial processes include melting of aluminum and recycling of metal scraps and used batteries, manufacturing of plastics, steel and galvanized pipes, chemicals, pharmaceuticals, cosmetics, cooking utensils, and roofing sheets. Residential estates are also found within the Ota industrial estate [5], [6].
Fine particulate matter (PM), i.e. particles with diameter less than 2.5 μm ($PM_{2.5}$), is a complex mixture of chemicals including loosely bound toxic metals that have been associated with several adverse health and environmental problems [1], [2]. Health risks resulting from exposure to trace metals include reproductive toxicity, cancer, genetic damage, and neurotoxicity [7], [8]. Several literatures have established that metals from anthropogenic sources are commonly found in the fine

particles. Industrial processes, coal combustion, construction and demolition, refuse incinerations, vehicle exhausts and non-exhaust sources such as wear from brakes and tyres are the major sources of Cd, Cr, Ni, Cu, Zn, Mn, Pb, V, Fe, Ba, K, and Al [9], [10]. Chemical speciation is an important analytical tool that offers a way for the elucidation of the chemical form(s) and the quantitative estimation of a specific element in toxicological and biochemical investigations to assess the bioavailability and mobility of metals in environmental and bio-physicochemical systems [10]-[15]. In order to evaluate the risk posed by metal toxicity, it is important to focus more on the determination of bio-available metals rather than total metals since they are readily available once inhaled [8], [10]. However, information on the pollution status of the present study area is sparse with nonexistent research data ever documented on the characterization and risk evaluation of atmospheric fine particulate bound metals. In addition, there are limited studies on the chemical forms of trace metals bound to atmospheric particulates in other cities in Nigeria. Therefore, the key objectives of this study are: (i) to determine and investigate the concentrations and partitioning of fine particulate matter bound trace metals (Cr, Cd, As, Ni, Zn, Cu and Mn); (ii) to evaluate the carcinogenic health risks and non-carcinogenic health risk posed to residents due to inhalation of particulate-bound metals in the study area.

2. Experimental

2.1. Sample collection

$PM_{2.5}$ samples were collected from the study sites on 47 mm diameter PTFE filters using Environtech Gravimetric Sampler (Model SLE-FPS105) between July 2014 and January 2015, exception of August. The filters were equilibrated in a desiccator for 48 hours and weighed thrice before and after sampling for accuracy, using four digits microbalance (Mettler Toledo Me 204). Filters were handled only with tweezers coated with Teflon tape to reduce the possibility of contamination. Laboratory and field blank filters were collected to reduce the gravimetric bias due to filter handling before, during and after sampling. A total of 20 samples were collected. After weighing, samples were preserved in the refrigerator at 4 ^oC prior to analysis.

2.2. Determination of chemical speciation

The modified Tessier derived procedure developed by [16], using a combination of mechanical shaker and micro wave oven was followed to fractionate the $PM_{2.5}$-bound trace metals into four fractions, namely, soluble and exchangeable metals (F1), acid soluble carbonates, Mn, Fe oxides and reducible metals (F2), metals bound to organic matter, oxidisable and sulphidic metals (F3) and residual metals (F4). Table 1 outlines the multistep procedure adopted for this work. Extracts from each fraction were carefully filtered and aliquot of 10 mL of the supernatant was drawn with a micropipette and placed in a preconditioned 15 mL polyethylene tube. The same four-step chemical extraction procedure was performed on blank filters. All the extracts were preserved at 4° C in the refrigerator before analysis with ICP-MS. Sequential extraction instrument used were, pH meter (mettler Toledo), mechanical shaker (ZWY-11OX 30, Zhicheng), high speed centrifuge (HC-3514, Zonkia), and closed vessel microwave reaction system (CEM MARS 6, One Touch Technology). All reagents used in this research were of analytical grade.

Table 1. Chemical speciation reagents and extraction conditions.

Metallic fraction	Reagent	Experimental conditions
F1: Soluble and exchangeable metals	15 mL H_2O (PH= 7.4)	Shaker agitation, 3 H at RT[a] 3 min centrifugation at 5000 rmp
F2: Carbonates, oxides & reducible metals	10 mL $NH_2OH.HCl$ (0.25M) at pH=2.0	Shaker, 2H at RT 3 min centrifugation at 5000 rmp
F3: Bound organic matter, oxidizable	a) 7.5 mL H_2O_2, 30% b) + 7.5 mL H_2O_2, 30%	Shaker at $\Delta T(95^oC)^b$ Shaker at $\Delta T(95^oC)^b$

and sulphidic metals	c)+15mL NH_4AcO [2.5M] at pH=3.0	Shaker,90 min at RT 3 min centrifugation at 5000 rmp
F4: Residual metals	10 mL (HNO_3: HCl: HF) (ratio = 2: 6:2) + boric acid	20 min at MW oven

[a]RT= room temperature; [b]ΔT= applied heat; MW= microwave [16]

2.3. Determination of bioavailability index (BI)

The first two fractions (F1 and F2) are labile fractions and completely bioavailable. The bioavailability index (BI) was calculated as shown in equation 1:

$$BI = F1 + F2 \Big/ \sum_{i=1}^{4} F_i \qquad (1)$$

2.4. Analysis of trace metals using inductively coupled plasma mass spectrometry

All solution samples were analyzed for 7 trace metals, Cr, Cd, As, Ni, Zn, Cu and Mn, using ICP-MS (Agilent 7500ce Series), equipped with an octopole collision cell following the instrument operating conditions. Calibration of the ICP-MS instrument was performed with multi-element calibration standards from 10 to 1000 mg/L. During the measurement, ^{6}Li, ^{45}Sc, ^{72}Ge, ^{103}Rh, ^{115}In, ^{159}Tb, ^{71}Lu and ^{209}Bi at 100 mg/L were used as the internal standards to monitor possible instrumental drifting and to compensate the matrix effect. The calibration of the ICP-MS instrument performed with multi-element calibration standards showed that the correlation coefficients (r^2) for all trace metals ranged from 0.994 to 1.000. Recovery test results done to evaluate the efficiency of the extraction procedure for the samples was in the range of 85% – 95%, method detection limits (MDLs) values for trace metals ranged from 4.29×10^{-5} to 2.56×10^{-3} μg/m^3, while the method precision, expressed as relative percent difference (RPD) was in the range of 5 to 12%. All quality control procedures were strictly adhered to from sampling to data analyses and report [17], [18].

2.5. Health risks assessment

The health risk estimate for a receptor exposed to the target trace metal species via the inhalation pathway was characterized into two forms namely non-carcinogenic and carcinogenic health risk. It was assessed using the inhalation dosimetry methodology outlined in EPA's Risk Assessment Guidance for Superfund (RAGS). The non-carcinogenic health risk, was computed by using the hazard quotient (HQ) formulae as follows:

$$Inhalation\ HQ = EC / (RfC \times CF) \qquad (2)$$

Where HQ is the hazard quotient, EC is the exposure concentration (μg/m^3), RfC is the inhalation reference concentration (mg/m^3) and CF is the conversion factor, 1000 (μg/mg) [19,20]. The concentration of the target metal in air (μg/m^3) was used as the exposure metric [20]. The exposure air concentration (μg/m^3), (EC), was calculated for each individual contaminant according to equation 3:

$$EC = \frac{CA \times ET \times EF \times ED}{AT} \qquad (3)$$

Where CA is the concentration of the metal in the air to which the person is exposed (μg/m^3), ET is the exposure time (hours/day), EF is the exposure frequency (days/year), ED is the duration exposure (years), and AT is the averaging time (hours).

On the other hand, the cumulative lifetime cancer risk was calculated as shown:

$$Risk_{inhalation} = \sum_{i=1}^{n} IUR_i \times EC_i \qquad (4)$$

Where, IUR is the inhalation unit risk (μg/m^3)$^{-1}$ and EC is the exposure air concentration (μg/m^3). IUR and R*f*C values were obtained from databases provided by the EPA's Integration Risk Information System (IRIS) and Provisional Peer Reviewed Toxicity Values (PPRTVs) [19], [21].

3. Results and discussion

3.1. Chemical speciation of $PM_{2.5}$-bound trace metals

The import of easily released trace metals can only be known from determination of bio-available concentrations which is very useful in risk assessment of metal toxicity [7], [9]. In Ota industrial estate, speciation results were obtained for Cr, Cd, As, Ni, Zn, Cu and Mn, (figure 1). Increased percentage distribution of soluble and exchangeable fraction (F1) obtained in OTE were Cr (46%) and Cu (32%) while Cd (67%) was dominant in the reducible fraction (F2). Metals present in the soluble and exchangeable fraction amongst all other fractions pose the greatest risk to human health because they are easily adsorbed on particle surfaces and released into aqueous solutions [13], [16]. This implies that Cr and Cu would be readily bioavailable once inhaled into the respiratory system. However, even though Cd and Ni is present, it is only dominant in the carbonates, oxides and reducible fractions. Therefore, Cd and Ni can only be a threat to human health under acidic aquatic conditions. The chemical speciation trend in the $PM_{2.5}$-bound trace metal levels was F4 > F2 > F3 > F1. Trace metals present in the residual fraction are said to be stable due to their stronger association with the crystalline structures of the minerals and as such are of no threat to the human health.

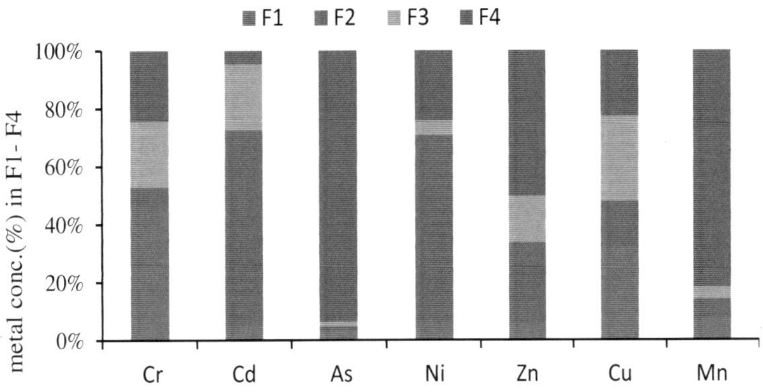

Figure 1. Average fractionation percentage of $PM_{2.5}$-bound trace metals.

3.2. Bioavailability index (BI)

Bioavailability index of metals shows the amount of soluble fraction present in the cell membrane through inhalation pathway [7]. The calculated trace metal bioavailability index results presented in table 2 indicated that Cd and Ni had the highest BI value of 0.7, followed by Cr and Cu (0.5). Amongst these metals Cd, Cr and Ni are the most toxic and their higher bioavailability index values indicate the possibility of their being fully available to physiological activities once inhaled into the respiratory system [14], [15].

Table 2. Bioavailability index of $PM_{2.5}$-bound trace metals.

Trace metals (μg/m^3)	Cr	Cd	As	Ni	Zn	Cu	Mn
Soluble fraction	4.78	0.449	0.232	0.111	32.8	0.173	0.054
Insoluble fraction	4.25	0.169	11.5	0.041	64.5	0.187	0.326
Bioavailability index	0.5	0.7	0.02	0.7	0.3	0.5	0.1

3.3. Carcinogenic health risk of inhaled $PM_{2.5}$-bound trace metals

Table 3 shows the incremental lifetime cancer risk (ILCR) at the study sites for four (4) carcinogenic trace metals, As, Cd, Cr and Ni capable of producing tumors of the respiratory tract, nasal passage, trachea, bronchus, and lung. It is usually assumed that even a minute exposure to these trace metals is a potential risk [8], [22]. The USEPA considers cancer risks between 1×10^{-6} (i.e., 1 in 1,000,000) and 1×10^{-4} (i.e., 1 in 10,000) to be generally acceptable [20]. Bioavailable form rather than total content of trace metal in airborne $PM_{2.5}$ was used in order to give a better and more accurate health risk estimate [7], [14], [15]. Two approaches were considered: cumulative risks to adult and children aged 0-6 and 6-18 years at the site due to inhalation of trace metals in air. From results presented in Table 3, the sum of the cumulative lifetime cancer risks across all investigated carcinogens due to inhalation exposures ranged as follows: adults (4.25×10^{-2}); children 1-6 years old (4.87×10^{-3}) and children 6-18 years old (1.46×10^{-2}). It is evident that the results obtained were above USEPA's acceptable risk range of 1×10^{-6} to 1×10^{-4}. Individual carcinogen showed that the range of the incremental lifetime cancer risks of Cd and Ni due to inhalation exposures for adults and children 1-6 and 6-18 were within the acceptable risk range of 6×10^{-4} to 6×10^{-2} μg/m^3 and 4×10^{-3} to $4.0\times10^{-}$ μg/m^3, respectively. On the contrary, Cr recorded elevated levels for adults (4.11×10^{-2} μg/m^3) and children within ages 6–18 (1.42×10^{-2} μg/m^3) above the USEPA's stipulated risk of 8.0×10^{-5} - 8.0×10^{-3} μg/m^3. This would result in greater than a one-in-ten-thousand increased risk of developing cancer during one's lifetime. Also from Table 3, As exceeded EPA's acceptable risk range of 2.0×10^{-4} - 2.0×10^{-2} μg/m^3. These results indicate that one (1) person out of a million people could get cancer after exposure to arsenic trace metals in ambient air inhaled from the investigated study area. Similar findings were observed by [15]. Thus, air particle pollution may pose serious health risks to the residents within the affected age categories in the study area.

3.4. Non-carcinogenic health risk of inhaled $PM_{2.5}$-bound trace metals

When an HQ and HI value exceeds one (HQ>1; HI>1), there is a possibility that some non-cancer effects may occur, the reverse happens with (HQ>1; HI>1) [19], [23]. The hazard quotients (HQs) of Cr, Cd, As, Ni and Mn were computed to determine the non-carcinogenic health risks for adults, children aged 0-6 and 6-18 years at the study site. Evaluation of the overall potential risk, by summing HQ values for the five trace metals gave the hazard index (HI). Results in Table 4, showed that the HQ for adults, children aged 0-6 and 6-18 years ranged between 2-80, 1-46 and 3-137, respectively. Implying a significant potential for non-cancer adverse health effects from all the five inhaled trace metals since the HQ is greater than 1 (HQ>1).

Table 3. Carcinogenic health risks assessment of $PM_{2.5}$-bound trace metals via inhalation.

Categories	Elements	CA	EC	IUR	$Risk_{inhalation}$	$\sum Risk_{inhalation}$
Adults	Cr	4.78	3.43	1.2×10^{-2}	4.11×10^{-2}	4.25×10^{-2}
	Cd	4.49×10^{-1}	3.22×10^{-1}	1.8×10^{-3}	6.00×10^{-3}	
	As	2.32×10^{-1}	1.67×10^{-1}	4.3×10^{-3}	7.00×10^{-4}	
	Ni	1.11×10^{-1}	8.00×10^{-2}	2.4×10^{-4}	1.91×10^{-5}	
Children 1-6 years	Cr	4.78	3.93	1.2×10^{-2}	4.72×10^{-3}	4.87×10^{-3}
	Cd	4.49×10^{-1}	3.69×10^{-2}	1.8×10^{-3}	6.64×10^{-5}	
	As	2.32×10^{-1}	1.91×10^{-2}	4.3×10^{-3}	8.21×10^{-5}	
	Ni	1.11×10^{-1}	9.12×10^{-3}	2.4×10^{-4}	2.19×10^{-6}	
Children 6-18 years	Cr	4.78	1.18	1.2×10^{-2}	1.42×10^{-2}	1.46×10^{-2}
	Cd	4.49×10^{-1}	1.12×10^{-1}	1.8×10^{-3}	2.02×10^{-4}	
	As	2.32×10^{-1}	5.73×10^{-2}	4.3×10^{-3}	2.46×10^{-4}	
	Ni	1.11×10^{-1}	2.74×10^{-2}	2.4×10^{-4}	6.58×10^{-6}	

Table 4. Non-carcinogenic health risks assessment of $PM_{2.5}$-bound trace metals via inhalation.

Categories	Elements	CA	EC	*RfC*	HQ	HI=∑ HQ
Adults	Cr	4.78	8.00	1.0×10^{-4}	80	194
	Cd	4.49×10^{-1}	7.52×10^{-1}	1.0×10^{-5}	75	
	As	2.32×10^{-1}	3.89×10^{-1}	1.5×10^{-5}	26	
	Ni	1.11×10^{-1}	1.86×10^{-1}	1.4×10^{-5}	13	
	Mn	5.37×10^{-2}	9.00×10^{-2}	5.0×10^{-5}	2	

Children 1-6 years	Cr	4.78	4.58	1.0×10^{-4}	46	111
	Cd	4.49×10^{-1}	4.30×10^{-1}	1.0×10^{-5}	43	
	As	2.32×10^{-1}	2.23×10^{-1}	1.5×10^{-5}	15	
	Ni	1.11×10^{-1}	1.06×10^{-1}	1.4×10^{-5}	8	
	Mn	5.37×10^{-2}	5.15×10^{-2}	5.0×10^{-5}	1	
	Cr	4.78	13.7	1.0×10^{-4}	137	334
Children 6-18 years	Cd	4.49×10^{-1}	1.29	1.0×10^{-5}	129	
	As	2.32×10^{-1}	6.68×10^{-1}	1.5×10^{-5}	44	
	Ni	1.11×10^{-1}	3.19×10^{-1}	1.4×10^{-5}	23	
	Mn	5.37×10^{-2}	1.54×10^{-1}	5.0×10^{-5}	3	

4. Conclusions

The highest concentrations for carcinogenic metals Cr and Cd were found in the soluble and exchangeable (F1) and reducible metals (F2) fractions of fine particles, respectively. The sum of the cumulative lifetime cancer risks due to inhalation exposures across all investigated carcinogens for adults, children 1-6 years old and children 6-18 years old were found above EPA's acceptable risk range while the cancer risks for each individual carcinogen indicated exceedance in only As and Cr trace metals. HQ and HI for all categories computed were found greater than one (HQ>1;HI>1) which is an indication of significant health risks most especially to residents and persons who by reason of their occupation spend most of their time within the study area. There is an urgent need for enforcement for pollution control and abatement policy and frequent industrial emission monitoring.

5. Acknowledgments

The authors gratefully thank, Prof. Zhang Suojiang, the Director of the Institute of Process Engineering, Chinese Academy of Sciences, Beijing, China, for providing necessary laboratory support. The study is supported by research grant from Covenant University, Nigeria.

6. References

[1] Anake W U, Ana, G R E E and Benson N U 2016a Study of surface morphology, elemental composition and sources of airborne fine particulate matter in Agbara industrial estate, Nigeria *Int. J.Appl Env Sci*, **11** 881-90

[2] Anake W U, Benson N U and Ana, GREE 2016b Characterization of airborne fine particulate matter ($PM_{2.5}$) and its air quality implications in Ogun State, Nigeria. *Proc. 3rd Covenant University Int. Conf. on African Development Issues* 543-47

[3] Obioh I B, Ezeh G C, Abiye O E, Alpha A, Ojo E O and Ganiyu A K 2013 Atmospheric particulate matter in Nigerian megacities *Toxicol Environ Chem* **95** 379-85

[4] Etim E U (2012). Estimation of pollution load from an industrial estate, South-western Nigeria *Afr J Environ Sci and Technol* **6** 125-9

[5] OSGN (Ogun State Government of Nigeria) 2008 The Ogun State regional development plan (2005- 2025). Abeokuta: Office of the Governor, Ogun State

[6] Onilude O O "Industrial pollution in Ota Ogun State, Nigeria 2015 The disconnect between citizen, industry and government perspectives", M.Sc. Thesis, Environmental Applied Science and Management Toronto Ontario, Canada

[7] Varshney P, Saini R and Taneja A 2016 Trace element concentration in fine particulate matter ($PM_{2.5}$) and their bioavailability in different microenvironments in Agra, India: a case study *Environl Geochem Health* **38** 593–605

[8] Khanna I, Khare M, and Gargava P 2015 Health Risks Associated with Heavy Metals in Fine Particulate Matter: A Case Study in Delhi City *India J. Geosci Environ Protec* **3** 72

[9] Mukhtar A and Limbeck A 2013 Recent developments in assessment of bio-accessible trace metal fractions in airborne particulate matter: A review. *Anal. Chimica Acta* **774**, 11- 25

[10] Adamu H, Luter L, Mohammed M L, and Umar B A 2013 Chemical speciation: A strategic pathway for insightful risk assessment and decision making for remediation of toxic metal contamination *Environ. Pollut* **2** 92-9

[11] Benson N U, Anake W U, Essien J P, Enyong P and Olajire AA 2016a Distribution and risk assessment of trace metals in *Leptodius exarata*, surface water and sediments from Douglas Creek, Qua Iboe estuary *J Taibah Univ Sci* http://dx.doi.org/10.1016/j.jtusci.2016.08.004

[12] Benson N U, Asuquo F E, Williams A B, Essien J P, Ekong CI, Akpabio O, Olajire A A 2016b Source evaluation and trace metal contamination in benthic sediments from equatorial ecosystems using multivariate statistical techniques *PLoS ONE* **11** e0156485. doi:10.1371/journal.pone.0156485

[13] Benson N U, Anake W U and Olanrewaju I O (2013). Analytical relevance of trace metal speciation in environmental and biophysicochemical systems. *Am J Analyt Chem*, **4**, 633-41.

[14] Betha R, Pradani M, Lestari P, Joshi U M, Reid J S and Balasubramanian R 2013 Chemical speciation of trace metals emitted from Indonesian peat fires for health risk assessment *Atmos. Res* **122** 571-8

[15] Feng X D, Dang Z, Huang W L and Yang C 2009 Chemical speciation of fine particle bound trace metals *Int. J. Environ. Sci.* **6** 337-46

[16] Fernandez-Espinosa A J, Rodríguez M T and Álvarez F F 2004 Source characterisation of fine urban particles by multivariate analysis of trace metals speciation *Atmos Environ.* **38** 873-86

[17] US EPA United States Environmental Protection Agency 1999 Compendium method for the determination of inorganic compounds in ambient air. EPA/625/R-96/010a. United States EPA, Office of Research and Development, Washington DC

[18] Ventura L M B, Amaral B S, Wanderley K B, Godoy J M and Gioda A 2014 Validation method to determine metals in atmospheric particulate matter by inductively coupled plasma optical emission spectrometry *J. Braz. Chem Soc.* **25** 1571-82

[19] IEc Industrial Economics Inc. 2009 Risk assessment of the inhalation of particulate matter from lake bed sediments in the upper Columbia river/lake Roosevelt

[20] US EPA United States Environmental Protection Agency 2009 Risk assessment guidance for superfund. vol. 1: human health evaluation manual (Part f, supplemental guidance for inhalation risk assessment); Office of Superfund Remediation and Technology Innovation U.S. Environmental Protection Agency: Washington D.C. EPA-540-R-070-002

[21] IRIS, Integrated Risk Information System 2015 A–Z List of substances Retrieved from http://cfpub.epa.gov/ncea/iris/index.cfm?fuseaction=iris

[22] US EPA (U.S. Environmental Protection Agency) 1989 Risk assessment guidance for super fund volume i human health evaluation manual Part a EPA/540/1-89/002

IOP Conf. Series: Earth and Environmental Science **68** (2017) 012006 doi:10.1088/1755-1315/68/1/012006

Modelling of Major Flood Arrivals on Chinese Rivers by Switch-time Processes

Pavel Stoynov[1], Plamena Zlateva[2], Dimiter Velev[3] and Xuejun Zong[4]

[1] FEBA, Sofia University "St. Kliment Ohridski", Sofia, Bulgaria
[2] Institute of Robotics, Bulgarian Academy of Sciences, Sofia, Bulgaria
[3] University of National and World Economy, Sofia, Bulgaria, Email: dgvelev@unwe.bg
[4] Shenyang University of Chemical Technology, Shenyang, China

Abstract. Nowadays, there is a considerable flood risk in China is. A brief description of major floods on Yangtze River and Huang He River is given. Both big Chinese rivers have long records of floods with severe life and property damages. Quantification of the stochastic behaviour of the largest floods is a key task in the risk assessment and mitigation. An exponential distribution of the time intervals between consecutive floods is assumed in classical study of inter-arrival times of floods. An approach for modelling of flood arrivals on both Chinese rivers by switch-time (ST) processes is proposed. These ST distributions can be considered as distributions of sums of random number exponentially distributed random variables. The proposed model specifies explicitly times of occurrence not only of floods but also of higher risk of potential floods. This approach could be useful for making prognoses of floods and for analysing changes in hydrologic behaviour of rivers.

1. Introduction

In recent years, the climate changes lead to an increase in both the number and severity of floods throughout the world. The floods can cause economic and social damage, and lead to serious consequences for the environment and critical infrastructure [1].

A simple definition of the flood is an overflow of a body of water (river, lake) that submerges land or the abnormal accumulation of water on the surface due to excess rainfall and rise of the groundwater level above surface on impermeable or saturated terrains. The major floods have the following characteristics: due to overflowing of rivers and lakes, unexpected and serious breaks in dikes, levees and other protective structures or uncontrolled releases of dam water; coverage of a wide continuous area and rapid spreading to adjacent areas of relatively lower elevation; relatively deep in most parts of the flood-stricken areas [2].

Since the beginning of registration of floods in the world, there are about fifteen major floods [3], [4]. It can be seen that the greatest number of these major floods have occurred in China. The 1931 Central China flood is considered as the deadliest disaster in the world. It started in July 1931, when Yellow river (Huang He), Huai River and Yangtze River water level heights are exceeded. Both big rivers Yangtze and Huang He have long records of floods with significant damages [5].

Large and damaging floods occur every year. The maximum estimated annual damage caused by river floods in one country is recorded in China in 2010, where a total loss of US $51 billion is reported [1]. For this reason, the each approach to the assessment of the flood risk in China is very important and useful.

2. Quantitative statement of the problem

Quantification of the stochastic behavior of the largest floods is a key task in flood risk assessment.

Usually it is assumed that floods occur independently with some average rate of occurrence [7]. For this reason, their occurrence can be modeled by Poisson process and inter-arrival times are independent and exponentially distributed [8]. In this case, they have the distribution of a random variable $\xi \in Exp(\beta)$ with probability density function.

$$f_{\xi}(x) = \begin{cases} \frac{1}{\beta} e^{-\frac{x}{\beta}}, x \geq 0 \\ 0, x < 0 \end{cases} \quad (1)$$

Therefore, the exponential distribution for intervals between two major floods is assumed [9].

Some observations on inter-arrival times however show that they are heavy-tailed. In probability theory, heavy-tailed distributions are probability distributions whose tails are heavier than the tails of the exponential distribution [8].

A typical representative of heavy-tailed distributions is Pareto distribution. It is applied in some investigations of times between consecutive floods. Heavy tails introduced with Pareto (power law) distribution for these intervals is proposed in [10].

General Pareto distribution $P(\mu, \sigma, \chi, \alpha)$ is a distribution with cumulative distribution function

$$F(x) = 1 - [1 + \left(\frac{x-\mu}{\sigma}\right)^{-\frac{1}{\chi}}]^{-\alpha}, \; x > \mu. \quad (2)$$

Here $\mu \in R$ is a centric parameter, $\sigma > 0$ is a scale parameter, $\chi > 0$ is inequality parameter, and $\alpha > 0$ is a form parameter. Different special cases of this distribution are used [8]

$$P_1(\sigma, \alpha) = P(\sigma, \sigma, 1, \alpha), \qquad P_2(\mu, \sigma, \alpha) = P(\mu, \sigma, 1, \alpha), \qquad P_3(\mu, \sigma, \chi) = P(\mu, \sigma, \chi, 1). \quad (3)$$

Heavy tailed distributions actually present the possibility for longer inter-arrival times. Another approach to present possibility for longer inter-arrival times is to skip some of the arrivals in the Poisson process. In this case, the inter-arrival time is with Erlang distribution. The Erlang distribution is a two parameter family of continuous probability distribution which may be considered as a sum of independent exponentially distributed random variables.

The approach proposed in this article is to randomize the Erlang distribution of inter-arrival times. This leads to so called switch-time (ST) distributions and switch-time (ST) processes. ST distributions can be considered as distributions of sums of random number exponentially distributed random variables. In this way, without using heavy tailed distributions it can be presented longer and more flexible inter-arrival times [11], [12]. An approach for modelling Inter-arrival time of floods by ST processes is proposed in [13].

The aim of this paper is to model flood arrivals on rivers by switch-time processes. In particular, the major floods of the both Chinese rivers Yangtze and Hunag are considered. These ST processes permit to represent both seasonal and cyclical risks of floods. A benefit of using this approach is that the model specifies explicitly times of occurrence not only of floods but also of higher risk for potential floods. With consistent data sets for flood appearance, the approach could be useful for forecasting of floods and analyzing of changes in hydrologic behaviour of rivers. The model can track also consequences of risk mitigation activities, for example building of dams.

3. Essence of ST distributions and ST processes

It is said that a random variable ξ with probability mass function $f_{\xi}(x)$ has distribution of $ST(n, \beta)$ family and denote this fact $\xi \in ST(n, \beta)$, if the probability mass function of ξ is given by the formula

$$f_{\xi}(x) = \begin{cases} \sum_{k=1}^{n+1} P(D^n = k) f_{\xi}(x \mid D^n = k) \\ 0, x < 0. \end{cases} = \begin{cases} \sum_{k=1}^{n+1} P(D^n = k) f_{G^k}(x), x \geq 0 \\ 0, x < 0. \end{cases} \quad (4)$$

Where G^k are random variables with probability mass function $f_{G^k}(x) = f(k,\beta)$ and D^n are positive integer mixing random variables.

Here for G^k it can be adopted different families of distribution.

In this article, the case of ST family of first kind is considered where:

$$G^k \in \Gamma(k,\frac{1}{\beta}) \equiv Erlang(k,\frac{1}{\beta}), \quad \text{i.e.} \quad \xi \,|\, D^n \equiv Erlang(D^n,\frac{1}{\beta}). \tag{5}$$

Correspondingly, D^n is a random variable, taking values $k = 1,\ldots,(n+1)$ with probabilities:

$$P(D^n = k) = \frac{C(n,\beta)n!}{\beta^k (n-k+1)!}, \quad k = 1,\ldots,(n+1) \tag{6}$$

Where the coefficients $C(n,\beta)$ are given by the formulas:

$$C(n,\beta) = \frac{1}{I(n,\beta)}, \qquad I(0,\beta) = \frac{1}{\beta}, \qquad I(n,\beta) = \frac{1}{\beta} + \frac{n}{\beta} I(n-1,\beta), \quad n = 1,2,\ldots \tag{7}$$

Also, variables $\tilde{D}^n = D^n - 1$ can be introduced taking values $k = 0,\ldots,n$ with probabilities

$$P(\tilde{D}^n = k) = \frac{C(n,\beta)n!}{\beta^{k+1}(n-k)!}, k = 0,\ldots,n. \tag{8}$$

Then the probability mass function $f_\xi(x)$ of ξ may be presented also by the formula

$$f_\xi(x) = \begin{cases} \sum_{k=0}^{n} P(\tilde{D}^n = k) f_\xi(x \,|\, \tilde{D}^n = k) \\ 0, x < 0. \end{cases} = \begin{cases} \sum_{k=0}^{n} P(\tilde{D}^n = k) f_{G^{k+1}}(x), x \geq 0 \\ 0, x < 0. \end{cases} \tag{9}$$

If a random variable has ST distribution of first kind, we will denote this fact $\xi \in ST1(n,\beta)$. For such a variable, the probability density is density of Erlang distribution with first parameter being another integer-valued random variable with a specific distribution.

If in $ST1(n,\beta)$ it is enforced $D^n = k$, i. e.

$$P(D^n = k) = 1, \; P(D^n = i) = 0, \qquad 1 \leq i \leq k-1 < k+1 \leq i+1 \leq n+1, \tag{10}$$

Which may be considered as a degenerate ST distribution of first kind, and it is actually obtained Erlang distribution, i.e. $\xi \in Erlang(k,\frac{1}{\beta})$.

The distribution $ST1(n,\beta)$ can be considered as a special kind of generalized Gamma distribution [12]. It is said that a random variable Λ has generalized gamma distribution if its probability density function is given by

$$f_\Lambda(x) = \frac{u^{s-\alpha}}{\Gamma(\alpha)U(\alpha,\alpha+1-s,u\beta)} e^{-x\beta} x^{\alpha-1}(u+x)^{-s} \tag{11}$$

where $x > 0$, $-\infty < s < +\infty$, $\alpha > 0$, $\beta > 0$, $u > 0$ and

$$U(a,b,z) = \frac{1}{\Gamma(a)} \int_0^\infty e^{-zt} t^{a-1} (1+t)^{b-a-1} dt, \qquad \text{where } a > 0, \; z > 0. \tag{12}$$

The above expression is the integral representation of the hyper-geometric function of second kind [14]. In this case, for Λ it is given $\Lambda \in G\Gamma(\alpha,\beta,u,s)$.

IOP Conf. Series: Earth and Environmental Science **68** (2017) 012006 doi:10.1088/1755-1315/68/1/012006

It needs to remember that the random variable ξ has a Gamma distribution and it is denotes as $\xi \in \Gamma(\alpha,\beta)$, if its probability mass function is given by

$$f_{\xi}(x) = \begin{cases} \dfrac{\beta^{\alpha}}{\Gamma(\alpha)} x^{\alpha-1} e^{-\beta x}, \; x \ge 0 \\ 0, \; x < 0 \end{cases}. \tag{13}$$

Here $\Gamma(\alpha)$ is defined by $\Gamma(\alpha) = \int_0^{\infty} x^{\alpha-1} e^{-x} dx.$ It is seen that $\Gamma(\alpha,\beta) = G\Gamma(\alpha,\beta,u,s=0).$

Actually, it can written that $ST1(n,\beta) = G\Gamma(\alpha=1,\beta,u=1,s=-n).$

The Exponential distribution is a special kind of $ST1(n,\beta)$ distribution and $ST1(0,\beta) \equiv Exp(\beta)$. To recall that the random variable ξ has exponential distribution and it is denotes as $\xi \in Exp(\beta)$, if its probability mass function is given by

$$f_{\xi}(x) = \begin{cases} \beta e^{-\beta x}, \; x \ge 0 \\ 0, \; x < 0. \end{cases} \tag{14}$$

The case $ST1(1,\beta) \equiv Lindley(\beta)$. can be considered as a weighted version of exponential distribution with weighting function $w(x) = 1 + x$. It can said that the random variable ξ has Lindley distribution and it is denotes as $\xi \in Lin(\beta)$, if its probability mass function is given by

$$f_{\xi}(x) = \begin{cases} \dfrac{\beta^2}{\beta+1}(1+x)e^{-\beta x}, \; x \ge 0 \\ 0, \; x < 0 \end{cases}. \tag{15}$$

The case $n = 2$ leads to $ST1(2,\beta)$ distribution which can be defined as weighted exponential distribution by the weight function $w(x) = (1+x)^2$.

It is said that the random variable ξ has $ST1(2,\beta)$ distribution and it is denotes as $\xi \in ST1(2,\beta)$, if its probability mass function is given by

$$f_{\xi}(x) = \begin{cases} \dfrac{\beta^3}{\beta^2 + 2\beta + 2} e^{-\beta x}(1+x)^2, \; x \ge 0 \\ 0, \; x < 0. \end{cases} \tag{16}$$

To check that the weight function in the definition of the distribution is the right one, it could calculate as $Ew(\xi) = \dfrac{\beta^2 + 2\beta + 2}{\beta^2}$.

It is said that a process $X(t)$ is a ST process of first kind or $ST1(n,\beta)$ process, and it is denoted this fact as $X(t) \in ST1(t;n,\beta)$, if the following statements are satisfied:

1) $X(0) = 0$.
2) $X(t)$ is pure jump process with jumps at times T_i $i = 1,2,\ldots$ and jump sizes $\Delta X(T_i) = 1$.
3) The intervals between two jumps are $\tau_i = T_i - T_{i-1} \in ST1(n-1,\beta)$, $i = 0,1,\ldots$ $T_0 = 0$

It is said that $X(t)$ is compound $ST1(n,\beta)$ process if it is replaced condition 3) with condition:

3') $X(t)$ is pure jump process with jumps at times T_i $i = 1,2,\ldots$ and jump sizes $\Delta X(T_i) = Y_i$, where Y_i are independent and identically distributed random variables.

4. Brief description of major Floods on Yangtze River and Huang He River

4.1. Description of major floods on Yangtze River

The Yangtze River (actually Chang Jiang, the Long River) is the longest river in Asia, the third longest in the world and the longest in the world to flow entirely within one country. It drains one-fifth of the land area of China and its river basin is a home to one-third of the country's population. Flooding along the river has been a major problem.

The rainy season in China is May and June in areas south of Yangtze River, and July and August in areas north of it. The huge river system receives water from both southern and northern flanks, which causes its flood season to extend from May to August. Meanwhile, the relatively dense population and rich cities along the river make the floods more deadly and costly. Tens of millions of people live in the floodplain of the Yangtze valley, an area that naturally floods every summer and is habitable only because it is protected by river dikes. The floods large enough to overflow the dikes have caused great distress to those who live and farm there. Floods of note include 1931, 1935, 1954, 1998.

The 1931 Central China floods were a series of floods. The floods are generally considered among the deadliest natural disasters ever recorded, and almost certainly the deadliest of the 20th century (when pandemics and famines are discounted). The 1931 floods covered more than 77 700 square km—including the cities of Nanjing and Wuhan—killed more than 300 000 people, and left 4 000 000 more homeless. The Yangtze again flooded in 1935, causing great losses of life. From June to September 1954, the Yangtze River floods were a series of catastrophic floorings that occurred mostly in Hubei Province. Due to unusually high volume of precipitation as well as an extraordinarily long rainy season in the middle stretch of the Yangtze River late in the spring of 1954, the river started to rise above its usual level in around late June. Despite efforts to open three important floodgates to alleviate the rising water by diverting it, the flood level continued to rise until it hit the historic high of 44.67 m in Jingzhou, Hubei and 29.73 m in Wuhan. The number of deaths from this flood is estimated at around 33 000, including those who died of plague in the aftermath of the disaster. The 1998 Yangtze River floods was a major flood that lasted from middle of June to the beginning of September 1998 in China at the Yangtze River. The event is considered the worst Northern China flood in 40 years. In the summer of 1998, China experienced massive flooding of parts of the Yangtze River, resulting in 3 704 dead, 15 million homeless and $26 billion in economic loss. Other sources report a total loss of 4 150 people, 180 million people are affected, and 13.3 million houses are damaged.

The floods caused $26 billion in damages. One of the major objectives of the Three Gorges Dam project was to alleviate flooding on the lower Yangtze. The dam proved effective during the extraordinarily rainy summer of 2010 by holding back much of the resultant floodwaters and thus minimizing the impact of flooding downstream. However, the dam still had to open its floodgates to reduce the high water volume in the reservoir, and flooding and landslides in the Yangtze basin killed several hundred people and caused extensive property damage [15].

4.2. Description of major floods on Huang He River

Huang He (The Yellow River) is the third-longest river in Asia, following the Yangtze River and Yenisei River, and the sixth longest in the world at the estimated length of 5464km.

There is a long record of floods caused by Huang He before 19th century. The September–October 1887 flood has been estimated to have killed between 900 000 and 2 million people, and is the second-worst natural disaster in history, excluding famines and epidemics. The Yellow River more or less adopted its present course during the 1897 flood. During the 1931 flood according to various estimates from 1 000 000 to 4 000 000 people are killed, and 88 000 square km are completely inundated, and approximately 21,000 square km more are partially flooded, leaving 80 million people homeless. It is the worst natural disaster recorded (excluding famines and epidemics) On 9 June 1938, during the Second Sino-Japanese War, Nationalist troops under Chiang Kai-Shek broke the levees holding back the river near the village of Huayuankou in Henan. The goal of the operation was to stop the advancing Japanese troops by following a strategy of "using water as a substitute for soldiers" (yishui daibing). The 1938 flood of an area covering 54 000 squared kilometer took some 500 000 to 900 000 Chinese lives, along with an unknown number of Japanese soldiers. Huang He floods in 1887, 1931 and 1938 collectively killed millions and are considered to be the three deadliest floods in history and among the most destructive natural disasters ever recorded.

Throughout most of its history, China has attempted to control the Huang He by building overflow channels and increasingly taller dikes, and in 1955 the Chinese embarked on an ambitious 50-year construction plan and flood-control program. This program included extensive dike construction, repair, and reinforcement, reforestation in the loess region, and the construction of a series of dams to control the river's flow, produce electricity, and supply water for irrigation. Silt-retaining dams have not been completely effective (the accumulation of silt reduces the power-generating capacity of the dams), and they have been criticized by environmentalists. Continued silting in the Huang He has remained a serious problem; however, the river has not burst its banks since 1945, in large part because of the flood-control program [16].

5. Applications of Switch-time processes for modelling of major flood arrivals in China

Applications of Switch-time processes for modelling of major flood arrivals on big Chinese rivers are proposed in the paper. In particular, both Rivers Yangtze and Huang He are considered.
In [16] an example with modelling inter-arrival time between floods by ST processes is presented.
Let us suppose that there are two types of events related to floods. The first type is "condition for floods". The process, which counts number of appearances of this kind of events, is Poisson with parameter β. The second kind of events we call "real flood". It can be modelled by the process $ST1(n=[\frac{t_{\max}}{t_{\min}}],\beta)$, where $t_{\max}$ is the maximal and $t_{\min}$ - the minimal observed interval between two major floods in a country (area, region) and $[.]$ is the integer part of a number.

For Yangtze river it is estimated that $\frac{1}{\beta}=35$ and $n=[\frac{t_{\max}}{t_{\min}}]=\frac{61}{12}=5$. The realizations of the random variables D_i^n are as shown in Table 1.

TABLE 1. Yangtze RIVER INTER-ARRIVAL TIME OF FLOODS MODELING

N	Year of major flood of Yangtze River	Time interval (years)	D_i^n
1	1870	61	5
2	1931	23	2
3	1954	44	4
4	1998	12	1
5	2010	-	-

This means that at every 35 years a possibility of major flood occurs and at every 5 possibilities it can be expected such a flood to really occur.
If we apply Poisson-process modelling, the interval between two consecutive floods should be with exponential distribution. In this case, the mean interval between two floods is 35 years and the model says that a flood should be expected every 35 years. It is seen that this model do not catch the variations of the inter-arrival interval which are better reflected by the model with ST processes.

For Huang He river it is estimated that $\frac{1}{\beta}=12{,}75$ and $n=[\frac{t_{\max}}{t_{\min}}]=\frac{26}{7}=3$. The realizations of the random variables D_i^n are as shown in Table 2.

This means that at every 13 years a possibility of major flood occurs and at every 3 possibilities it can be expected such a flood to really occur.
If we apply Poisson-process modelling, the interval between two consecutive floods should be with exponential distribution. In this case, the mean interval between two floods is 13 years and the model says that a flood should be expected every 13 years. It is seen that this model do not catch the variations of the inter-arrival interval which are better reflected by the model with ST processes.

When we model with ST processes, the parameter β is related to the average interval between times of high risk of flood when the conditions for (potential) flood are available.
The applied approach gives the possibility to "calibrate" the severity of the floods captured in the model. If we select different thresholds for a flood to be considered as major, we may obtain different degree of refinement of the flood's scale.
For the considered scales of floods of Yangtze River and Huang He River, we may say that conditions for large-scale flood of Huang He River appear more frequently than conditions for large-scale flood of Yangtze River.

TABLE 2. Huang He River Inter-arrival Time of Floods Modeling

N	Year of major flood of Huang He River	Time interval (years)	D_i^n
1	1887	10	1
2	1897	8	1
3	1905	26	3
4	1931	7	1
5	1938	-	-

6. Conclusions

An approach for modelling of flood arrivals on rivers by switch-time processes is proposed. In particular the major floods of the both Chinese rivers Yangtze and Hunag are considered. These ST processes permit to represent both seasonal and cyclical risks of floods. ST processes can be used to model floods even for short data records. With longer periods covered and more precise information about floods the models using ST processes could give a better picture of inter-arrival times.
A benefit of using this approach is that the model specifies explicitly times of occurrence not only of floods but also of higher risk for (potential) floods. With consistent data sets for flood appearance, the approach could be useful for making prognoses of floods and analyzing changes in hydrologic behaviour of rivers. The model can track also consequences of risk mitigation activities, for example building of dams.

7. Acknowledgments

The authors express their gratitude to the Bulgarian National Science Fund for the support under the Grant № DFNI-I02/15 from 2014 and the Bulgarian – Chinese scientific cooperation Grant № DNTS/ China 01/6 from 2014.

8. References

[1] Z.W Kundzewicz, et al., 2014 Flood risk and climate change: global and regional perspectives. *Hydrological Sciences Journal*, **59** (1), pp 1–28.
[2] PreventionWeb-Serving the information needs of the disaster reduction community, Hazard – Foods, [Online] Available: http://www.preventionweb.net/english/hazards/flood/
[3] Rankred 2015 *Internet-Technology_Knowledge*, 15Biggest and Worst Floods Ever in History, Nature, Available: http://www.rankred.com/top-10-biggest-and-worst-floods-ever-in-history
[4] O'Connor J and Costa E 2004 *The World's Largest Floods, Past and Present: Their Causes and Magnitudes*. U. S. Geological Survey Circular, **1254**.
[5] List of deadliest floods, Wikipedia, the free encyclopedia, Available: https://en.wikipedia.org/wiki/List_of_deadliest_floods
[6] Asian Development Bank. *Water-related disasters and disaster risk management in the People's Republic of China*, Mandaluyong City, Philippines: ADB, 2015.
[7] Singh V P, Wang S X, and Zhang L 2005 Frequency analysis of nonidentically distributed hydrologic flood data, *J. Hydrol.*, **307**, pp.175–195.

[8] Stoynov P 2006 *Manual of statistics*. Sofia University "St. Kliment Ohridski", Sofia.
[9] Enzel Y, Redmond K, House K and Biondi F 2002 *Climate variability and flood frequency at* decadal *to millenial time scales*, Annual Meeting, Geol. Soc. of Am., Denver, Colorado.
[10] Benson D, Schumer R and Meerschaert M 2007 Recurrence of extreme events with power-law interarrival times, *Geophys. Res. Lett.*, **34**, L16404, doi: 10.1029/2007GL030767.
[11] Stoynov P 2010 Negative binomial distribution and applications. *Proc. of the Third Int. Conf. on Financial and Actuarial Mathematics*, Sofia, Bulgaria, pp 2-20.
[12] Stoynov P 2013 Stoynov Process–Definition and Simulations. *Proc. of the Sixth Int. Conf. on Financial and Actuarial Mathematics*, Sofia, Bulgaria, pp 97-104.
[13] Stoynov P, Zlateva P and Velev D 2015 An Approach for Modelling Inter-arrival Time of Floods *Int. Journal of Innovation, Management and Technology*, **6** (4), pp 267-271.
[14] Zamani H and Ismail N 2010 Negative binomial-Lindley distribution and its applications *Journal of Mathematics and Statistics*, **6** (1), pp 4-9.
[15] Encyclopedia *Britannica: Yangtze River* 2016 http://www.britannica.com/eb/article-9110538/Yangtze-River
[16] Encyclopedia *Britannica: Huang He River* 2016 http://www.britannica.com/place/Huang-He.

Characterisation of Polychlorinated Biphenyls in Coastal Inland Seawater, Nigeria

Nsikak U Benson, Omowunmi H Fred-Ahmadu, Adebusayo E Adedapo, Winifred U Anake

Analytical and Environmental Chemistry, Unit, Department of Chemistry, Covenant University, Km 10 Idiroko Road, Ota, Ogun State, Nigeria.

E-mail: nsikak.benson@covenantuniversity.edu.ng

Abstract. The concentrations of polychlorinated biphenyls (PCBs) were determined in samples of surface seawater collected from five (5) different locations along the coastline of a tropical lagoon. The surface water samples were analysed using gas chromatography coupled with electron capture detector (GC-ECD) with fifteen (15) PCB congeners detected across the studied area. The chlorobiphenyls levels are in the order Ebute Ero > Oko Baba > Ijora > Lagos Island > Unilag lagoon front with levels 107.89, 53.15, 34.90, 27.78, and 2.15 mg/L, respectively. The enhanced average concentration of PCBs at the Ebute Ero site is due to the predominance of PCB-180 and PCB-185. A negative correlation was found between the total PCB concentration and the level of dissolved oxygen. Although, the results indicated fairly high levels of PCBs, the anthropogenic contributions from industrial releases and domestic activities may be largely associated with the detected concentrations of the di-, tetra-, penta-, hexa-, hepta-, and octa-chlorobiphenyls.

1. Introduction

Over the years, there have been heightened concerns over the profound toxicity and potential human and ecosystem risks associated with the presence of polychlorinated biphenyls (PCBs) in the environment. PCBs are man-made organic environmental pollutants made up of carbon, hydrogen and chlorine atoms, and are generally known to be persistent, bioaccumulative and toxic (PBT) chemicals [1]-[5]. The level of toxicity exhibited by each PCB congener is a function of the pattern of chlorine substitution. Coplanar PCBs, particularly (PCB-77, 81, 126 and 169) and mono-ortho PCBs (PCB-105, 114, 118, 123, 156, 157, 167 and 189) are the most toxic congeners [6].

They are high boiling; chemically stable and non-flammable compounds hence their application as coolants, electrical insulators, plasticizers, transformer lubricants, and in capacitors etc. PCBs enter the environment during their manufacture process, use, and disposal. They emerge from accidental spills and leaks during their transport, and from leaks from disposed capacitors and transformer or burning of PCB products. Exposure to PCBs is known to cause skin rashes, acne, and liver damage [7]. They have been classified as probable carcinogens by the USEPA. Although the use of PCBs was banned in the US and most European countries in the 1970s, they still enter the marine environment through disposed PCB-containing materials, disposal of obsolete electrical equipment, and industrial discharge [8]. Due to their slow biodegradability, they settle into seawater and sediments resulting in increased load of highly chlorinated PCBs for the deep-water habitat [7], [9]. The role of floating plastic resin microparticulates in the transportation of contaminants within the marine environment resulting in their accumulation in the surface microlayer has been reported [10]. This may pose a considerable toxic risk to the marine ecosystem and commercial fishing.

In Nigeria, coastal lagoons and lacustrine ecosystems have witnessed increasing physical, biological and chemical pollutions that emanate mostly from human mediated sources [11]-[13]. The major anthropogenic PCBs input into water resources include biomass burning, leakages from transformers and shipwreck, and illegal disposal of PCB-laden end-of-life electronic items [14]. In aquatic ecosystems, humans can be exposed to PCBs through oral ingestion, dermal contact and inhalation. Surface water constitutes a primary pathway transportation of particulate bound and water loosely absorbed PCBs. Thus, the concentration of PCBs in aquatic ecosystem is largely influenced by the quantity of the particulates suspended in the water column that actually adsorbs chlorobiphenyls [3], [15]. This study presents the concentration of PCBs in surface water collected from Lagos lagoon, a tropical inland aquatic ecosystem.

2. Materials and methods

2.1. Study area

The study area is the Lagos lagoon which lies between latitude 6° 27' N and longitude 3° 23' E with a maximum length of 50 km, and about 13 km width. The lagoon is located in the Southwestern part of Lagos State, Nigeria. The Lagos lagoon has large surface area of about 6,354.7 km^2 and extends to various parts of the Lagos Island. It is mainly plied by small boats and canoes, which are used as means of commercial transportation and fishing. Activities carried out on the lagoon and along its coast include fishing, manufacturing plants, wood constructions, transportation, saw milling, plank markets, other markets and hundreds of wooden shacks. Apart from the domestic and economic activities carried out on the lagoon, large amount effluents and other pollutants such as nutrient runoff from agricultural lands, sewages, wood preservatives and waste disposed of nearby markets are discharged into the lagoon regularly. The upsurge of industrial effluent in addition to the various discharges has resulted in intense pollution of the lagoon, enhancing health risks posed to humans and the seawater environment. The coordinates of the sampling locations were taken using a Garmin-GPSMAP 76S handheld global positioning system (GPS). The specific sampled sites are Ebute Ero, EBE, (6°28'10" N, 3°23'04" E), Ijora, IJR, (6°27'55" N, 3°22'36" E), Lagos Island, LGL, (6°27'47" N, 3°23'01" E), Oko Baba, OKB, (6°28'50" N, 3°23'29" E), and the University of Lagos lagoon front, UWF, (6°30'18" N, 3°24'01" E).

2.2. Sample collection and treatment

Sampling was carried out using a boat. Surface seawater samples were collected using clean, and properly labelled amber glass bottles; physicochemical parameters were measured at the sampling sites using a pre-calibrated QPM – 7000 multi-probe analyser. Then samples were immediately stored in an ice-packed cooler and transported to the laboratory at the National Institute of Oceanography and Marine Research, Nigeria. They were then transferred into a refrigerator and stored at 4°C.

Sample extraction was carried out 2 days after sampling using liquid-liquid extraction. 100 mL of water sample was introduced into a separatory funnel and 50 mL of solvent, dichloromethane (DCM) was added. The mixture was shaken vigorously and the stopper removed at intervals to release pressure from the funnel. The extract was the carefully drained into a conical flask and the process was repeated the second time with 50 mL of DCM [16]. This extraction was done in batches and the solvent extracts are concentrated using a rotary evaporator. The clean up of extracts was done using column chromatography in accordance with US EPA method 3630B [17]. The 20 mL glass column was packed with glass wool to about 5 mL and activated silica gel (60-200 mesh size) from about 5 mL to 15 mL of the glass column. 2 g of baked sodium sulphate (Na_2SO_4) was used to cap the sample; this was done to remove any trace of water present in the sample. The glass column was then conditioned with 20 mL of DCM to ensure that the silica gel is properly packed. The DCM was discarded and the concentrated sample was introduced into the glass column and eluted with 60 mL of DCM. The eluent was collected and concentrated using the rotary evaporator to about 2 to 3 mL. The concentrated sample (1μL) was then injected into the Perkin Elmer Autosystem XL Gas Chromatograph HP 5890 Series II, equipped with an electron capture detector (ECD) for analysis.

2.3. Quality assurance/control
Analytical quality control included the strict adherence to documented procedures, laboratory standard operating procedures and calibration of analytical instrument. The GC-ECD was calibrated using high quality analytical standards from the U.S. Environmental Protection Agency (US EPA). Standards were run to check for the column performance, peak height, resolution, and detection limit (LOD). The LOD were defined as 3 times the signal-to-noise ratio (S/N). Results from recovery studies for the chlorobiphenyls indicated recovery efficiency between 94% and 106%.

3. Results and discussion
The physicochemical properties and concentrations of PCBs determined in surface water samples collected from the Lagos lagoon are reported in Tables 1 and 2, respectively. The physicochemical parameters measured across the five sampling locations did not show large variations as presented in Table 1. The salinity values ranged between 20.1 and 23.6 PSU, while pH, conductivity, temperature and dissolved oxygen values ranged between 7.96 and 8.18, 32.6-37.3 μS, 30.1-30.6°C, and 4.22-5.57%, respectively. A negative Pearson product-moment correlation, r (-5.71) was observed between the total concentrations of PCBs and dissolved oxygen (DO). This implies that higher concentrations of PCBs in seawater may likely contribute to the depletion of the amount of dissolved oxygen available for the biota in seawater environment. This trend may threaten the existence of sea creatures and plants. On the other hand, the salinity and conductivity values had a strong positive linear relationship (r = 0.99) as already established in literature. UWF recorded the lowest values of physicochemical parameters except for temperature (31°C) and dissolved oxygen (5.13).

Table 1: Physicochemical properties of sampled seawater.

Properties	EBE	IJR	LGL	OKB	UWF
Salinity (PSU)	23.4	23.3	23.6	22.2	20.1
pH	8.05	8.10	8.18	8.02	7.96
Conductivity (μS)	36.7	36.7	37.3	35.0	32.6
Temperature (°C)	30.2	30.1	30.1	30.6	31.0
Dissolved Oxygen (%)	4.22	5.57	4.32	4.45	5.13

Table 2: Mean concentrations of PCB congeners in the seawater samples (mg/L).

PCB Congener	IUPAC name	EBE	IJR	LGL	OKB	UWF
PCB-8	2,4′-DiCB	2.69	1.16	ND	ND	ND
PCB-60	2,3,4,4′-TetraCB	ND	ND	ND	5.64	ND
PCB-81	3,4,4′,5-TetraCB	ND	1.26	4.46	ND	ND
PCB-105	2,3,3′,4,4′-PentaCB	5.34	1.35	ND	3.73	2.15
PCB-114	2,3,4,4′,5-PentaCB	9.98	1.38	ND	1.68	ND
PCB-126	3,3′,4,4′,5-PentaCB	ND	1.71	8.70	ND	ND
PCB-128	2,2′,3,3′,4,4′-HexaCB	ND	2.35	4.38	8.44	ND
PCB-138	2,2′,3,4,4′,5′-HexaCB	ND	2.52	3.19	9.48	ND
PCB-153	2,2′,4,4′,5,5′-HexaCB	ND	6.88	3.89	ND	ND
PCB-156	2,3,3′,4,4′,5-HexaCB	ND	7.58	3.17	3.13	ND
PCB-170	2,2′,3,3′,4,4′,5-HeptaCB	2.77	8.71	ND	ND	ND
PCB-180	2,2′,3,4,4′,5,5′-	25.46	ND	ND	ND	ND

	HeptaCB					
PCB-185	2,2′,3,4,5,5′,6-HeptaCB	56.25	ND	ND	ND	ND
PCB-189	2,3,3′,4,4′,5,5′-HeptaCB	ND	ND	ND	ND	ND
PCB-195	2,2′,3,3′,4,4′,5,6-OctaCB	5.40	ND	ND	21.05	ND
∑PCBs		107.89	34.90	27.78	53.15	2.15

The total concentrations of PCBs ranged between 2.15 and 107.89 mg/L. Table 2 indicates that total concentration of chlorobiphenyls in the lagoon system followed the sequence Ebute Ero > Oko Baba > Ijora > Lagos Island > UniLagos lagoon front with levels 107.89, 53.15, 34.90, 27.78, and 2.15 mg/L, respectively. The UWF site had the lowest concentration due to the presence of PCB-105 as the only detected congener. This low concentration may be due to minimal pollution of the site being a relaxation spot for the University of Lagos community. No visible economic or industrial activities occur around the lagoon. The high average concentration of PCBs at the EBE site is due to the predominance of PCB-180 (25.46 mg/L) and PCB-185 (56.25 mg/L). However, the enhanced levels of these heptachlorobiphenyls at this may be attributed to the fact that the coastline of EBE is characterised by a beehive of heavy anthropogenic activities such as municipal waste burning, indiscriminate disposal of solid and liquid waste, industrial discharge, human faeces, and runoff from nearby markets into the sea water. In addition, the location also recorded the concentration of two pentachlorobiphenyls PCB-105 (5.34 mg/L) and PCB-114 (9.98 mg/L), which may suggest contributions from industrial and domestic sources. PCB-105 was the most prominent congener as it was detected in four of the sampled sites except LGL, followed by PCB-114, PCB-128, PCB-138 and PCB-156, which were detected in three sampled locations.

4. Conclusion

The concentration of PCBs in the Lagos lagoon was assessed by collecting samples from 5 different locations along the seawater course. PCBs were detected in all samples. The physicochemical properties of the seawater across the sampled locations were similar. A negative correlation was observed between the concentration of PCBs and the amount of Dissolved oxygen which indicates that PCB accumulation depletes DO levels in the sea water. Unilag water front recorded low level of PCBs while Ebute Ero sample had the highest concentrations of PCBs. It is recommended that human activities contributing to PCB accumulation in the seawater should be seriously controlled by environmental regulatory agency along the coastline, while a continuous monitoring program is instituted and conducted periodically to ensure that the low profile concentrations of PCBs in the lagoon system are kept within threshold levels.

5. Acknowledgments

The study is supported by research grant from Covenant University, Nigeria. The authors acknowledge Dr. J.P. Unyimadu of Nigerian Institute of Oceanography and Marine Research for analytical assistance.

6. References

[1] EPA 2017 February 13 Polychlorinated Biphenyls (PCBs). Retrieved March 29, 2017 from USEPA: https://www.epa.gov/pcbs/learn-about-polychlorinated-biphenyls-pcbs

[2] Gioia R, Lohmann J, Dachs C, Temme S, Lakaschus D, Schulz-Bull I, Hand and Jones K C 2008 Polychlorinated biphenyls in air and water of the North Atlantic and Arctic Ocean *J. Geophys. Res.* **113**, D19302, doi: 10.1029/2007JD009750

[3] IARC Working Group on the Evaluation of Carcinogenic Risk to Humans Polychlorinated biphenyls and polybrominated biphenyls. Lyon (FR): International Agency for Research on

Cancer 2016 (IARC Monographs on the Evaluation of Carcinogenic Risks to Humans, No. 107) 1 Exposure Data. Available from: https://www.ncbi.nlm.nih.gov/books/NBK361688/

[4] Khaled A, El Nemr A, Said T O, El-Sikaily A and Abd-Alla A M A 2004 Polychlorinated biphenyls and chlorinated pesticides in mussels from the Egyptian red sea. *Chemosphere*, **54** (10), 1407–1412

[5] Ragab S, El Sikaily A and El Nemr A 2016 Concentrations and sources of pesticides and PCBs in surficial sediments of the red sea coast, Egypt. *The Egyptian J. of Aquatic Res.* **42**, 365–374

[6] Giesy J P, Kannan K, Blankenship A L, Jones P D and Hilscherova K 2000 Dixon-like and non-dioxin-like toxic effects of polychlorinated biphenyls (PCBs): Implications for risk assessment *Central European J. of Pub. Health*, **8** (SUPPL.), 43–45

[7] ATSDR 2014 Polychlorinated biphenyls ToxFAQ Retrieved March 29, 2017, from Agency for Toxic Substances and Disease Registry: https://www.atsdr.cdc.gov/toxfaqs/tfacts17.pdf

[8] Webster L, Russell M, Walsham P, Hussy I, Lacaze J P, Phillips L, ... Moffat, C F 2014 Halogenated persistent organic pollutants in relation to trophic level in deep sea fish *Marine Pollution Bulletin* **88**(1–2), 14–27

[9] Mormede S and Davies I M 2003 Horizontal and vertical distribution of organic contaminants in deep-sea fish species *Chemosphere*, **50**(4) 563–574.

[10] Mato Y, Isobe T, Takada H, Kanehiro H, Ohtake C and Kaminuma T 2001 Plastic resin pellets as a transport medium for toxic chemicals in the marine environment. *Environmental Science and Technology* **35**(2), 318–324

[11] Benson N U, Adedapo A E, Eritobor A L and Udosen E 2015 Total dissolved inorganic carbon and physicochemical characteristics of surface microlayer and upper mixed layer water from Lagos Lagoon Nigeria *Global NEST J.* **17** (2), 334-343

[12] Benson N U, Essien J P, Asuquo F E and Eritobor A L 2014 Occurrence and distribution of polycyclic aromatic hydrocarbons in surface microlayer and subsurface seawater of Lagos Lagoon, Nigeria *Env. Monitoring and Assessment* **186** 5519–5529

[13] Williams A B and Benson N U 2010 Interseasonal hydrological characteristics and variabilities in surface water of tropical estuarine ecosystems within Niger Delta, Nigeria, *Env. Monitoring and Assessment* **165** 399-406

[14] UNEP 2009 Sustainable Innovation and Technology Transfer Industrial Sector Studies: Recycling – From E-waste to Resources. United Nations Environment Programme. Available fromhttp://www.unep.org/pdf/Recycling_From_e-waste_to_resources.

[15] Song X, Li J, Chen L, Cai Z, Liao C, Peng H and Xiong, H 2012 Determination of polychlorinated biphenyls in seawater using headspace solid-phase microextraction coupled with gas chromatography-mass spectrometry with the aid of experimental design. *J. of the Braz. Chem. Soc.* **23**(1) 132-141

[16] EPA 2008 Method 1668B: Chlorinated biphenyl congeners in water, soil, sediment, biosolids, and tissue by HRGC/HRMS. EPA-821-R-08–020. Washington (DC): Office of Water, Office of Science and Technology, Engineering and Analysis Division. United States Environmental Protection Agency

[17] US Environmental Protection Agency US EPA 1996 Method 3630 Revision B. Silica gel cleanup SW-846 Manual. Washington DC USA.

Modeling of CO_2 Solubility in Aqueous Potassium Lysinate Solutions at Post-Combustion CO_2 Capture Conditions

Y Bian and S Shen

School of Chemical and Pharmaceutical Engineering, Hebei University of Science and Technology, Shijiazhuang 050018, China

Abstract. Aqueous potassium lysinate (LysK) has been proposed as an alternative to aqueous alkanolamines for CO_2 capture due to fast kinetics and large absorption capacity. However, thermodynamic modeling for aqueous LysK system has not been available yet. In this work, a modified Kent-Eisenberg model with correlated equilibrium constants was developed to interpret the vapor-liquid equilibrium (VLE) data at postcombustion capture conditions. The predictions from the developed model are in good agreement with the experimental results with AAD within 19 %.

1. Introduction

Chemical absorption with amine-based solvents particularly monoethanolamine (MEA) is a proven technology for CO_2 capture process [1]. However, these solvents have several major drawbacks such as high regeneration energy, significant solvent loss and thermal and oxidative degradation [2]. Aqueous amino acid salts (AAS) are promising alternatives to conventional amine-based absorbents due to several good properties such as low volatility, low toxicity, resistance to oxidative degradation as well as fast kinetics with CO_2 [3]. In recently years, the extensive studies for several AAS systems have been reported [3]-[5]. The available solubility information for aqueous AAS, such as potassium/sodium salts of taurine, glycine, proline and sarcosine, was summarized and reported in the recent literature [6]-[9]. In our previous work, we have studied the reaction kinetics of CO_2 absorption into aqueous potassium solution of basic amino acids including lysine in a wetted wall column [3]. It was observed that LysK shows faster kinetics with CO_2 than the industrial standard MEA. Moreover, experimental data on CO_2 solubility in aqueous LysK for concentrations (1.25–3.28 $mol\cdot kg^{-1}$) at post-combustion CO_2 capture conditions have been recently reported in open literature [10], [11]. However, thermodynamic modeling for aqueous LysK system has not been available yet.

LysK possesses a primarily α-amino group and an aliphatic amino group in its side chain, which can exist in different dissociation states dependant on the pH of solution. As mentioned in the literature, the pH of fresh LysK solution is usually in the range of 11 – 13, anionic state of lysine is expected (Figure 1). Apart from the primarily amino group, the amino group in side chain might also participate in reactive absorption in these basic environments [3]. The lack of experimental equilibrium constants of carbamate hydrolysis makes the modeling challenging.

In the present work, a modified Kent-Eisenberg model with correlated equilibrium constants was developed to interpret the vapor-liquid equilibrium (VLE) data at postcombustion CO_2 capture conditions.

$$\underset{LysH^{2+}}{^{+}H_3N\text{-}(CH_2)_4CH(NH_3^+)COOH} \leftrightarrow \underset{LysH^{+}}{^{+}H_3N\text{-}(CH_2)_4CH(NH_3^+)COO^-} \leftrightarrow \underset{LysH}{^{+}H_3N\text{-}(CH_2)_4CH(NH_2)COO^-} \leftrightarrow \underset{Lys^{-}}{H_2N\text{-}(CH_2)_4CH(NH_2)COO^-}$$

Figure 1. Dissociation equilibria of lysine in aqueous solutions

2. Modified kent-eisenberg model

Equilibrium reactions involved in the chemical absorption of CO_2 in aqueous LysK solution can be described as follows [6], [7]:

$$^{+}H_3N\text{-}R_2R_1CH\text{-}NHCOO^{-} + H_2O \xrightleftharpoons{K_1} {}^{+}H_3N\text{-}R_2R_1CH\text{-}NH_2 + HCO_3^{-} \quad (1)$$

$$^{-}OOCHN\text{-}R_2R_1CH\text{-}NH_2 + H_2O \xrightleftharpoons{K_2} H_2N\text{-}R_2R_1CH\text{-}NH_2 + HCO_3^{-} \quad (2)$$

$$LysH^{+} \xrightleftharpoons{K_3} LysH + H^{+} \quad (3)$$

$$LysH \xrightleftharpoons{K_4} Lys^{-} + H^{+} \quad (4)$$

$$CO_2 + H_2O \xrightleftharpoons{K_5} HCO_3^{-} + H^{+} \quad (5)$$

$$HCO_3^{-} \xrightleftharpoons{K_6} CO_3^{2-} + H^{+} \quad (6)$$

$$H_2O \xrightleftharpoons{K_7} H^{+} + OH^{-} \quad (7)$$

Where $R_1 = -COO^{-}$, $R_2 = -(CH_2)_4-$ and B denotes the bases in the AAS solution.
Equilibrium constants for reactions (1)-(7):

$$K_1 = \frac{[LysH][HCO_3^{-}]}{[R_2LysR_1COO^{-}]} \quad (8)$$

$$K_2 = \frac{[Lys^{-}][HCO_3^{-}]}{[^{-}OOCR_2LysR_1]} \quad (9)$$

$$K_3 = \frac{[LysH][H^{+}]}{[LysH^{+}]} \quad (10)$$

$$K_4 = \frac{[Lys^{-}][H^{+}]}{[LysH]} \quad (11)$$

$$K_5 = \frac{[HCO_3^{-}][H^{+}]}{[CO_2]} \quad (12)$$

$$K_6 = \frac{[CO_3^{2-}][H^{+}]}{HCO_3^{-}} \quad (13)$$

$$K_7 = [H^{+}][OH^{-}] \quad (14)$$

The pH of LysK solutions is generally greater than 8.0, thus we assumed that $LysH^{2+}$ can be neglected. Mass balance equations and charge balance equations for various species in LysK solutions are given as below:

AAS mass balance:

$$[K^{+}] = [LysH]_0 = [LysH] + [LysH^{+}] + [Lys^{-}] + [^{-}OOCR_2LysR_1] + [R_2LysR_1COO^{-}] \quad (15)$$

CO_2 mass balance:

$$\alpha[LysK]_0 = [CO_2] + [^{-}OOCR_2LysR_1] + [R_2LysR_1COO^{-}] + [CO_3^{2-}] + [HCO_3^{-}] \quad (16)$$

Charge balance:

$$[K^{+}] + [H^{+}] + [LysH^{+}] = [Lys^{-}] + 2[^{-}OOCR_2LysR_1] + [R_2LysR_1COO^{-}] + 2[CO_3^{2-}] + [HCO_3^{-}] + [OH^{-}] \quad (17)$$

The concentration of molecular CO_2 in the solution was obtained using Henry's law relationship:

$$P_{CO_2} = H_{CO_2}[CO_2] \quad (18)$$

Henry's constants in LysK solutions were obtained from our previous work [12]. The temperature dependence of the equilibrium constants for K_5-K_7 are well established and have been utilized in many VLE models [6]-[9]. They can be expressed as:

$$\ln K_i = C_1 + C_2 / T + C_3 \ln T \tag{19}$$

where C_1–C_3 are constants, which were taken from literature [6].

For several AAS solutions, the Kent-Eisenberg model with computational simplicity has been extensively used and presented good accuracy. Although the equilibrium constants (K_i) should not be composition dependent, the activity coefficients are. In the present model, the non-idealities have been lumped into the apparent constants $K_{i,app}$. For each data set, the unknown protonation and carbamation equilibrium constants in LysK solutions are obtained by fitting to experimental data using MATLAB software. The objective function F_{obj} for minimization was used in this study.

$$F_{obj} = \left(\frac{p_{CO2}^{Exp} - p_{CO2}^{Cal}}{p_{CO2}^{Exp}} \right)^2 \tag{20}$$

The apparent constants are assumed to be a function of temperature (T), CO_2 loading (α) and AAS concentration (C) [9], and the correlation is given to be:

$$\ln K_{i,app} = C_1 + C_2 / T + C_3 \ln T + b_1 \ln \alpha + b_2 / \alpha + b_3 / \alpha^2 + b_4 C_{\mathrm{LysK}} \tag{21}$$

Where C_1–C_3 and b_1–b_4 are fitting parameters.

3. Results and discussion

The available solubility data of CO_2 in aqueous LysK systems were used as the database to determine the constants in the modified Kent-Eisenberg model. The values of experimental CO_2 solubility reported in our previous work are listed in Table 1[11].

Table 1. Experimental solubility values of CO_2 for 2.317M LysK solution [a].

323.2 K		333.2 K		343.2 K	
α	P_{CO2} (kPa)	α	P_{CO2} (kPa)	α	P_{CO2} (kPa)
0.800	0.21	0.730	0.27	0.673	0.30
0.842	0.35	0.754	0.34	0.701	0.44
0.869	0.67	0.775	0.45	0.784	0.65
0.939	1.31	0.856	0.85	0.820	0.90
0.951	3.46	0.866	1.00	0.835	1.32
0.984	5.31	0.924	1.40	0.873	1.80
1.013	7.60	0.937	1.90	0.889	2.50
1.056	16.67	0.952	2.23	0.894	3.38
1.071	20.52	0.954	3.96	0.919	3.80
		0.988	9.45	0.942	4.28
		1.042	12.34	0.957	5.40
		1.027	15.14	0.959	7.30
		1.040	16.85	0.968	7.50
		1.045	19.20	0.985	10.56
				0.987	15.88

[a] α is the mole of carbon dioxide per mole of LysK, P_{CO2} is the equilibrium partial pressure of carbon dioxide. The standard uncertainties u are $u(T)$ = 0.1 K, u(α) = 0.012, $u(P_{CO2})$ =1.20 kPa.

The regressed values of equilibrium constants for amino-acid protonation and amino-acid carbamate formation are given as follows:

$$\ln K_1 = -716.799 + 28922.65T + 95.8767 \ln T + 49.1574 \ln \alpha + 93.2616/\alpha - 21.4617/\alpha^2 - 0.28889C \tag{22}$$

$$\ln K_2 = 2399.34 - 108177.0/T - 343.728 \ln T - 53.5652 \ln \alpha - 103.4309/\alpha + 23.9503/\alpha^2 + 0.16228C \tag{23}$$

$$\ln K_3 = 1286.12 - 39894.9/T - 120.105 \ln T - 370.162 \ln \alpha - 611.797/\alpha + 124.408/\alpha^2 - 0.03723C \tag{24}$$

$$\ln K_4 = -703.392 + 27646.5/T + 89.3781 \ln T + 51.5723 \ln \alpha + 99.6297/\alpha - 23.0394/\alpha^2 - 0.21364C \tag{25}$$

Where T is the temperature in K, α in moles of CO_2 per mole of LysK, and C in kmol m^{-3}.

Figure 2 shows the experimental VLE data for 2.317 M LysK solutions along with the model predictions from the modified Kent-Eisenberg model at different temperatures. The model demonstrates a reasonable representation of the experimental data used in the analysis. The average absolute relative deviation (AAD) between the experimental CO_2 partial pressures and those calculated from the model is 18.9%. The parity plot of both the measured and calculated CO_2 partial pressures shows insignificant deviation, as shown in Figure 3.

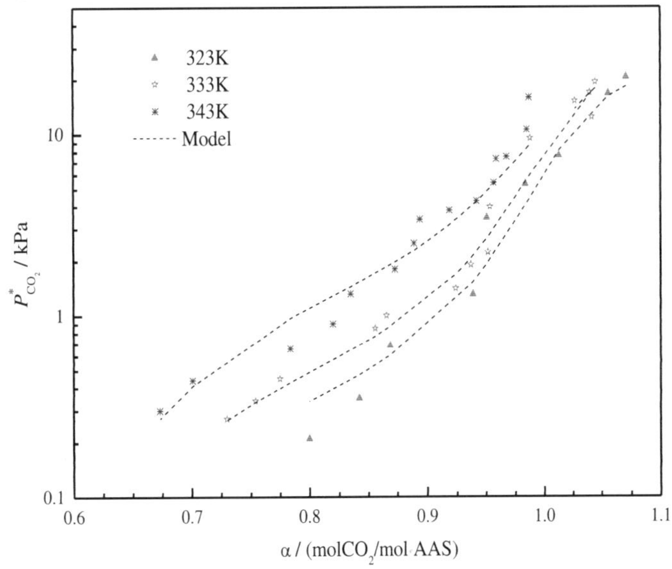

Figure 2. Comparison of CO_2 solubility for 2.317 M LysK solutions between experimental and predictions from the modified Kent-Eisenberg model.

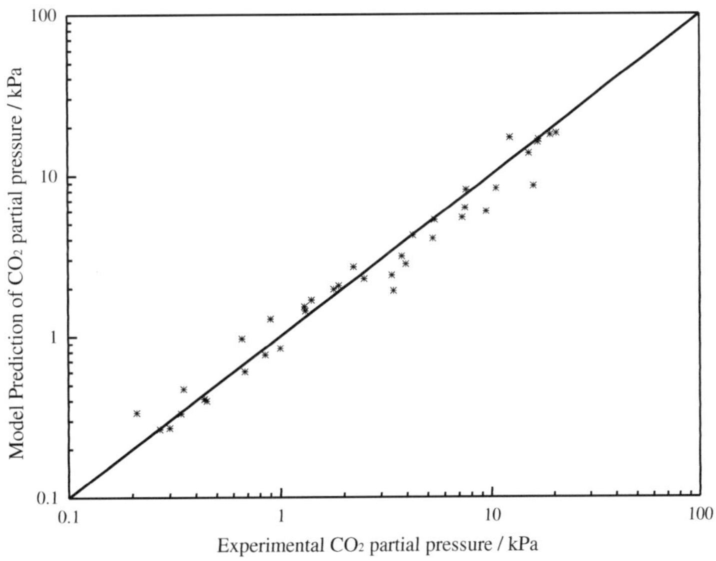

Figure 3. Parity plots between the experimental and the predicted from the modified Kent-Eisenberg model.

It should be noted that this model was obtained based on the limited database in the open literature. A robust VLE database for a wide range of temperature and CO_2 partial pressure for various LysK concentrations is demanding for better representation with the modified Kent-Eisenberg model in future work.

4. Conclusions

The vapor-liquid equilibrium data for aqueous LysK solutions can be well presented by a modified Kent-Eisenberg model. Equilibrium constants for amino-acid protonation and amino-acid carbamate formation were correlated as a function of temperature, CO_2 loading and concentration. The AAD between the experimental and the predictions is within 19 %. However, more robust VLE data points need to be determined in future work.

5. Acknowledgement

The authors would like to acknowledge Hebei Provincial Natural Science Foundation for Distinguished Young Scholars of China (B2015208067), Hebei Provincial Science and Technology Research Project of College and University (QN2015070) and Postgraduate Innovation Fund Project of Hebei Province (CXZZSS2017085) for financial support.

6. References

[1] Rochelle GT 2009 *Science* **325** 1652–54.
[2] Mumford KA, Wu Y, Smith KH and Stevens GW 2015 *Front. Chem. Sci. Eng.* **9** 125–41.
[3] Shen S, Yang Y, Bian Y and Zhao Y 2016 *Environ. Sci. Technol.* **50** 2054–63.
[4] Kumar PS, Hogendoorn JA and Versteeg GF 2003 *AIChE J.* **49** 203–13.
[5] Paul S and Thomsen K 2012 *Int. J. Greenhouse Gas Control* **8** 169–79.
[6] Majchrowicz ME, Kersten S and Brilman W 2012 *Chem. Eng. Sci.* **72** 35–44.
[7] Kumar PS, Hogendoorn JA, Timmer SJ, Feron PHM and Versteeg GF 2003 *Ind. Eng. Chem. Res.* **42** 2841–52.
[8] Hamzehie M and Najibi H 2015 *J. Nat. Gas Sci. Eng.* **27** 676–85.
[9] Mondal BK, Bandyopadhyay SS and Samanta AN 2015 *Int. J. Greenh. Gas Control* **36** 153–60.
[10] Shen S, Yang Y, Wang Y, Ren S, Han J and Chen A 2015 *Fluid Phase Equilib.* **399** 40–9.
[11] Zhao Y, Shen S, Bian Y, Yang Y and Ghosh U 2017 *J. Chem. Thermodynamics* **111** 100–5.
[12] Bian Y, Shen S, Zhao Y and Yang Y 2016 *J. Chem. Eng. Data* **61** 2391–8.

IOP Conf. Series: Earth and Environmental Science **68** (2017) 012009 doi:10.1088/1755-1315/68/1/012009

Optimization of high performance composite flocculant PFZSSB

K Zhang, SQ Li, Z Yan

School of Metallurgical and Ecological Engineering, University of Science and Technology Beijing, Beijing 100083, China

Abstract: Flocculant PFZSSB (MD-F02) was used to treat coking wastewater. In order to solve the problems such as excessive dosage and low flocculation effect in practical application, the dosing sequence and dosing time in the preparation of MD-F02 were studied to improve the flocculation effect. Based on a lot experiments, the optimal parameters were determined i.e. the dosing sequence of Zn> Fe> B, the dosing time of Zn, Fe, B of 5, 20, 10 min, respectively. The removal rate of SS was up to 90% and the removal rate of chemical oxygen demand (COD) was close to 50% under the dosage of modified MD-F02 of 125 mg/L.

1. Introduction

Coking wastewater is one of the most important pollution sources of water pollution. Its composition is complex and varied. It contains phenols, polycyclic aromatic compounds and heterocyclic compounds such as nitrogen, oxygen and sulfur. It is a typical industrial wastewater containing refractory organic compounds [1]. After the treatment of coking wastewater, the content of phenolic pollutants in effluent can be up to standard, but COD, NH_3-N and other indicators cannot reach the discharge standard [2]-[4]. Therefore, it is of great significance to reduce the pollution of coking wastewater and improve the utilization rate of wastewater.

Flocculation is one of the most widely used and most economical methods in wastewater treatment. Flocculant is core of flocculation technology. Advanced treatment of coking wastewater needs a cheap, high efficiency and simple process flocculant.

Zhou Jing et al. prepared composite flocculant polymeric aluminum silicate boron (PFASSB-MD-F01) by adding Fe, Al and B elements in the polymerization of silicic acid. Coking wastewater was treated by PFASSB, aluminium polychlorid (PAC) and polymeric aluminum sulfate (PAS). The results showed that PFASSB had a higher removal rate of SS and COD.

Yan Zhe et al. [5] made use of sodium silicate, ferric sulfate, zinc sulfate and sodium borate to prepare PFZSSB-MD-F02, and it was applied to the deep treatment of coking wastewater. When the dosage of flocculant was 175mg/L, SS removal rate of coking wastewater was more than 90%, the removal rate of COD was close to 50%, and the floc was larger than that of the same kind of flocculant. The effluent reached the national discharge level B standard.

In the process of coking wastewater treatment, MD-F02 has a lot of problems, such as high dosage and residue. In order to solve these problems, this paper optimized the adding sequence of Zn, Fe and B in the process of flocculant MD-F02 configuration, so as to achieve better treatment effect on coking wastewater and reduce the dosage of flocculant. Through the exploration of its mechanism, it provides some practical guidance and theoretical basis for the advanced treatment of coking wastewater, the development of flocculant and the development of the whole water treatment industry.

2. Experimental materials and methods

2.1. Experimental materials

The wastewater was collected from a Steel Corp after biochemical coking wastewater, contents of COD_{Cr} were 146 mg/L, contents of SS were 252 mg/L, pH=7.3.
The simulated wastewater was prepared with kaolin and sodium humic acid, and the content of SS was 100 mg/L, pH=7.3
Reagent: sulfuric acid, sodium silicate, ferric sulfate, zinc sulfate and sodium borate are all pure chemical reagents of Chinese medicine group.

2.2. Experimental methods

2.2. 1. Preparation of flocculant MD-F02.

The sodium silicate solution with SiO_2 mass fraction of 3% content was prepared. The sodium silicate solution was adjusted to a certain pH value by adding concentrated sulfuric acid. And then sodium silicate solution would be activated with a certain speed of about 60 min. A certain proportion of ferric sulfate, zinc sulfate and sodium borate were added to the activated silicic acid solution, and the stirring reaction was about 60 min. Static curing about 24 h, that is flocculant MD-F02.

2.2. 2. Change of pH value of solution.

In the process of preparing flocculant, the pH value of the solution was recorded after adding $ZnSO_4\cdot 7H_2O$, $Fe_2(SO_4)_3$ and $Na_2B_4O_7\cdot 10H_2O$, respectively. The pH value of the solution was continuously measured by PHSJ-3F type pH meter. The initial pH value of activated silicic acid was 2.5 and the reaction temperature was T=30 ℃.

2.2. 3. Detection methods of SS and COD.

A certain amount of flocculant was added into the wastewater. First, the solution was rapidly stirred for 60 seconds at 300 rpm. The solution was then slowly stirred at 30 rpm for 15 minutes. The solution was allowed to stand for 15 minutes and the contents of SS and COD in the supernatant were measured. SS in sewage using national standard method of measurement-Gravimetric method. The rapid digestion spectrophotometry was used for the determination of COD.

3. Results and discussion

3.1. The pH value of the reaction system was changed by adding Zn, Fe and B

3.1.1. The change of pH value in the activation of poly silicic acid.

Figure 1(a) shows the pH value of solution increased with the increase of Si dissolution time. Sodium silicate produces H_4SiO_4 during activation. In acidic solution, the silicic acid molecules combine with H^+ to form positive ion H_5SiO^{4+}, and the content increases with the decrease of pH value. Then, the silicic acid molecule reacts with the silicic acid monovalent cation to produce the polymerized silicic acid, thereby forming a plurality of monomer. In faintly acid and alkaline solutions, the polymerization of silicic acid with monovalent anions occurs. The OH^- generated in this reaction causes an increase in pH value. Part of the resulting dimer is further polymerized with OH^-.
The generated double silicate anion and silicic acid molecules can continue to participate in the polymerization reaction in accordance with the above reaction. This reaction will produce higher polymerization state of silicate anions and acid molecules. The highly polymerized silicic acid molecule is an important component of the active ingredient of the flocculant. From the whole process, H^+ is consumed continuously along with the activation of silicic acid, which leads to the increase of pH value.

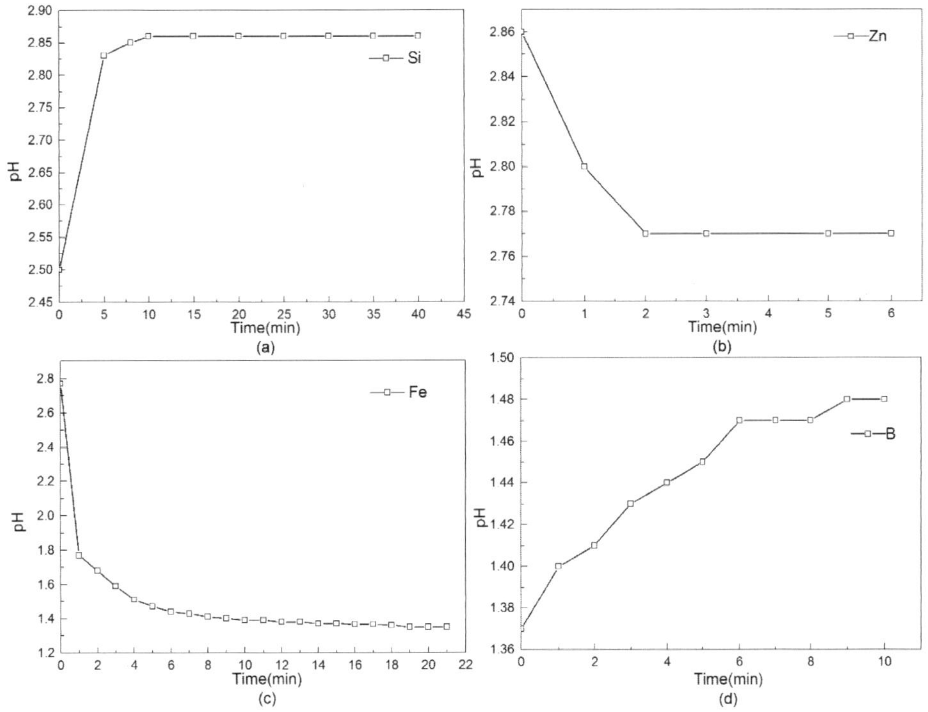

Figure 1 Change of pH value of solution.

3.1.2. Effect of Zn dissolution on pH value of silicic acid system.

Figure 1(b) shows the pH value of solution decreased with the increase of Zn dissolution time. Zinc sulfate is easily soluble in water, and the hydrolysis of zinc causes the solution pH to decrease:

$$[Zn(H_2O)_4]^{2+} \leftrightarrow [Zn(OH)(H_2O)_3]^{+} + H^{+} \tag{1}$$

Hydroxyl bridging interactions between hydroxyl complexes polymerize mononuclear complexes to form binuclear or multinuclear complexes:

$$2[Zn(OH)(H_2O)_3]^{+} \leftrightarrow [Zn_2(OH)_2(H_2O)_4]^{2+} + 2H_2O \tag{2}$$

$$[Zn_2(OH)_2(H_2O)_4]^{2+} + [Zn(OH)(H_2O)_3]^{+} \leftrightarrow [Zn_3(OH)_4(H_2O)_4]^{2+} + 2H_2O + H^{+} \tag{3}$$

At the same time, the complex will be hydrolyzed:

$$[Zn_2(OH)_2(H_2O)_4]^{2+} \leftrightarrow [Zn_2(OH)_3(H_2O)_3]^{+} + H^{+} \tag{4}$$

$$[Zn_3(OH)_4(H_2O)_4]^{2+} \leftrightarrow [Zn_3(OH)_5(H_2O)_3]^{+} + H^{+} \tag{5}$$

According to the hydrolysis process of zinc, with the continuous hydrolysis of zinc, H^+ was produced at the same time of the formation of polynuclear hydroxyl complex, which led to the decrease of pH value. At the same time, the polynuclear hydroxyl complex produced by the hydrolysis of zinc ions and the silicic acid skeleton were polymerized, which enriched the silicate framework. The pH value of the solution was stable after 5 minutes of zinc addition. Zinc hydrolysis is completed, so the best time for zinc dissolution of 5 minutes.

3.1.3. Effect of Fe dissolution on pH value of silicic acid system.

Figure 1(c) shows the pH value of solution decreased with the increase of Fe dissolution time. Iron (III) salt is ionized in aqueous solution to form Fe^{3+} ion, and Fe^{3+} exists in the form of hydrated ion $Fe(H_2O)_6^{3+}$. The reaction as follows:

$$Fe(H_2O)_6^{3+} + H_2O \rightleftharpoons Fe(H_2O)_5OH^{2+} + H_3O^{+} \tag{6}$$

The hydrolysis behavior of iron salt in aqueous solution is similar to that of zinc salt, and the final result is the decrease of pH value. The pH value of the solution was stable after 20 minutes. The hydrolysis of iron is complete, so the optimal time for iron dissolution is 20 minutes.

3.1.4. Effect of B dissolution on pH value of silicic acid system.
Figure 1(d) shows the pH value of solution increased with the increase of B dissolution time. B and Si have the similar chemical properties that the electronic acceptance is very strong. When an electron pair donor is present in the reaction environment, B can form a stable coordination compound. Boric acid is produced under strong acid conditions:

$$5H_2O + 2H^+ + B_4O_7^{2-} \rightarrow 4H_3BO_3 \quad (7)$$

Boric acid is a weak acid and its sodium salt is a strong base weak acid salt. Its aqueous solution is strongly alkaline. This explains why the pH value increases in the process of boron dissolution. The pH value of the solution was stable after 10 minutes of boron addition. The hydrolysis of boron is complete, so the best dissolving time is 10 minutes.

3.2. Effects of Zn, Fe and B dosing sequence on flocculation efficiency
According to the law of hydrolysis of Zn^{2+} and Fe^{3+}, the equilibrium of dissolution and precipitation of Zn^{2+} and Fe^{3+} exists in the solution. For the insoluble electrolytes such as $Zn(OH)_2$ and $Fe(OH)_3$, the change of the pH value of the solution will inevitably lead to the change of precipitation-dissolution equilibrium, eventually leading to the occurrence of dissolution or precipitation. From the point of view of the dissolution equilibrium, the hydrolysis polymerization process of Zn^{2+} and Fe^{3+} starts from the minimum value of pH that sol precipitation was found. $Zn(OH)_2$ and $Fe(OH)_3$ have the following dissolution precipitation equilibrium.

$$Fe(OH)_3 \Leftrightarrow Fe^{3+} + 3OH^- \quad (8)$$

$$K_{sp}\left(Fe(OH)_3\right) = \left[Fe^{3+}\right] \times \left[OH^-\right]^3 = 4.0 \times 10^{-38} \quad (9)$$

$$Zn(OH)_2 \Leftrightarrow Zn^{2+} + 2OH^- \quad (10)$$

$$K_{sp}\left(Zn(OH)_2\right) = \left[Zn^{2+}\right] \times \left[OH^-\right]^2 = 1.2 \times 10^{-17} \quad (11)$$

$$H^+ + OH^- \Leftrightarrow H_2O \quad (12)$$

$$K_w = 1 \times 10^{-14} \quad (13)$$

Assuming that the content of zinc ion and iron ion in the solution is 1mol/L, it is obtained from (9), (11):

$$[OH]^-_{Fe(OH)_3} = 3.42 \times 10^{-13} \quad (14)$$

$$[OH]^-_{Zn(OH)_2} = 3.46 \times 10^{-9} \quad (15)$$

$$pH_{Fe(OH)_3} = 1.54 \quad (16)$$

$$pH_{Zn(OH)_2} = 5.54 \quad (17)$$

The difference between these two values indicates that the affinity of Fe^{3+} to OH^-ion is much stronger than that of Zn^{2+} on OH^- ion, which means that Fe^{3+} is more prone to hydrolysis polymerization than Zn^{2+}. When Fe^{3+} and Zn^{2+} exist in the solution, the first hydrolysis polymerization is Fe^{3+}, and then with the increase of the pH value of the solution, the hydrolysis polymerization of Zn^{2+} occurs. In order to achieve more effective Zn, Fe complex with silicic acid skeleton, the effect of adding order of Zn and Fe on flocculation effect of MD-F02 should be taken into account.

The addition of sodium borate can cause the rapid rise of pH in silicate solution, leading to the occurrence of gel. Therefore, it is necessary to join the B element with Zn, Fe at the same time or after the Zn, Fe to join, in order to avoid the introduction of B element caused drastic changes in the solution pH value.

In order to investigate the effect of changing the dosing sequence on the flocculation effect, the following five dosing sequence schemes were set up:

Figure 2 shows that the order of addition play a key effect on the flocculation of MD-F02. Through the comparison of flocculation effect, the best dosing sequence is Zn>Fe>B. The reason is that iron elements can significantly reduce the pH of the solution system, compared to the effect of zinc and boron on the pH of the solution system. The pH value can affect the degree of polymerization of silicic acid molecules, the decrease of pH value can reduce the degree of polymerization of silicic acid molecules, weaken the bridging ability of MD-F02, and then influence the flocculation effect of flocculant.

3.3. Effect of optimized MD-F02 on coking wastewater treatment

Figure 3 shows the treatment effect of the modified flocculant MD-F022 is much higher than MD-F021. When the dosage of MD-F022 was 50 mg/L, the removal rate of SS was up to more than 90%. When the dosage of was 125 mg/L, the floc was the largest, and the removal rate of COD was up to 47%. The results show that the order of addition of Zn> Fe> B is more favorable for the polymerization of elements and poly silicic acid.

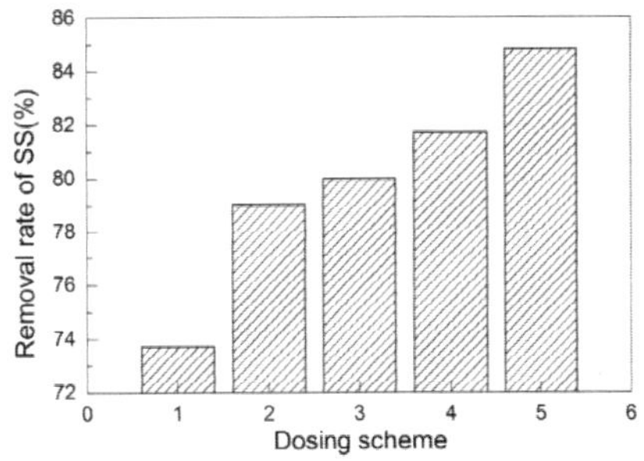

Figure 2. Flocculation effect of different dosing schemes.

(1# Fe=B=Zn; 2# Fe>B>Zn; 3# Fe>Zn>B; 4# Zn>B>Fe; 5# Zn>Fe>B)

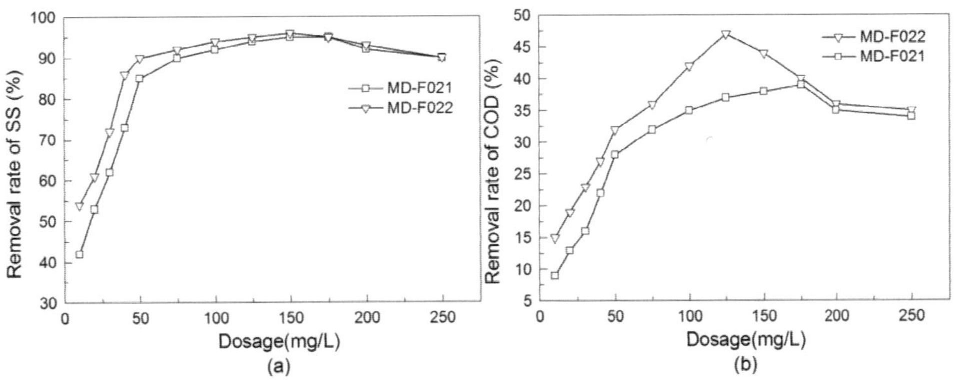

Figure 3. Comparison of coking wastewater treatment.

4. Conclusions

The effects of the addition of iron, zinc and boron on the pH value of the reaction system and the effect of the dosage sequence on the flocculation effect in the synthesis of MD-F02 were studied. The following conclusions were summarized:

(1) The dosage and time of Zn, Fe and B played a key effect on the flocculation efficiency of MD-F02, the best dosing sequence was Zn>Fe>B, and the corresponding optimal dissolution time was 5 min, 20 min, 10 min.

(2) When the dosage of modified flocculant MD-F02 was 125 mg/L, the SS removal rate of coking wastewater was above 90%, and the removal rate of COD was close to 50%, and the effluent reached the first grade B standard.

5. References

[1] X Ma, X Wang, Y Liu, et al 2017 Variations in toxicity of semi-coking wastewater treatment processes and their toxicity prediction. *Ecotoxicology and Environmental Safety*, **138**: 163-169.
[2] R Zhang, H You and Wu D 2016 Advanced treatment of coking wastewater by heterogeneous photo-Fenton technology with Cu/Fe oxide catalysts. *Desalination and Water Treatment*, **57**(26): 12010-12018.
[3] L Xue, J Liu, M Li, et al 2017 Enhanced treatment of coking wastewater containing phenol, pyridine, and quinoline by integration of an E-Fenton process into biological treatment. *Environmental Science and Pollution Research*, **24**(10): 9765-9775.
[4] X Jin, E Li, S Lu, et al 2013 Coking wastewater treatment for industrial reuse purpose: combining biological processes with ultrafiltration, nanofiltration and reverse osmosis. *Journal of Environmental Sciences*. **25**(8), 1565.
[5] Z Yan, SQ Li, XJ Hong, et al 2016 Preparation of efficient composite flocculant PFZSSB and its application to coke wastewater. *Chin. J. Proc. Eng*. (06), 1058.

Stimulation of Indigenous Carbonate Precipitating Bacteria for Ground Improvement

Adharsh Rajasekar[1], Charles K.S. Moy[1], Stephen Wilkinson[2]

[1]Department of Civil Engineering, Xi'an Jiaotong-Liverpool University, Suzhou 215123, Jiangsu, China.
[2]Department of Civil Engineering, University of Wolverhampton, WV1 1LY, United Kingdom.

Abstract. Calcite minerals are precipitated in soil through biomineralisation which can be either organic or inorganic in nature. Biomineralisation can be employed to improve ground conditions in its natural state. Usually, studies of applied biomineralisation are highly interdisciplinary involving expertise from engineers, chemists and microbiologists. In this paper, we study the potential of biomineralisation from indigenous bacteria present in soil. The soil samples were collected from a high permeable zone and the bacteria that inhabit the soil were stimulated at a temperature of 15°C. A cementation solution consisting of 500mM calcium chloride, urea and nutrient broth at a pH of 7.5 was added to the soil samples. Inorganic precipitation was found to be dominant and was more efficient when compared to organic precipitation. Carbonate precipitation data indicated that inorganic precipitation were 1.37 times better at carbonate formation in comparison to organic precipitation. Scanning Electron Microscopy analysis identified cementation bonds formed between soil particles. It was deducted that organic precipitation is dependent on temperature, and may take an extended time at such low temperature. The preliminary data presented in this paper suggests that the implementation of biomineralisation with in-situ microbes is promising but requires further laboratory and field investigation before being considered for engineering application.

1. Introduction

Bacterial application towards engineering purposes is attracting the attention of microbiologist and geotechnical engineers worldwide. Biomineralisation has recently gained popularity among researchers for its applicability towards ground improvement applications. Microbial activity that alters the chemical environment favoring mineral formation is known as Biomineralization [1], [2]. Microbially induced calcite precipitation (MICP) is a biochemical mechanism which is driven by the microorganism upon interacting with a chemical solution rich in calcium. Research in MICP has shown that microbially released CO_2 interacts with the biomineralization solution favoring carbonate formation. The carbonate combines with the calcium ion (Ca^{2+}) leading to the precipitation of calcium carbonate.

In the biomineralization process, bacteria commonly serve as nucleation sites for the precipitation of calcium carbonate. One of the important enzymes that's been shown to mediate high percentage of carbonates is urease. Urease (UA; EC 3.5.1.5) is a nickel-containing metalloenzyme found in a wide range of microorganisms [3]. Presence of urease initiates a process commonly known as ureolytic activity which aides in the conversion of urea to ammonia and CO_2 (Equation 1). Ureolytic activity promotes superior precipitation of carbonate (Equation 2 and 3). Ca^{2+} ions bind to the negatively charged bacteria surfaces, thus creating a neutral environment for Ca^{2+} adsorption (Equation 4).

Bacterial cells are very important for the precipitation of CaCO3, because the bacteria provide nucleation sites (heterogeneous nucleation) and affect the specific types of minerals formed (Figure 2). Okwadha and Li [4] found that high concentration of bacterial cells increases the amount of calcite precipitation by MICP, which happens because of the increase in the urease concentration for urea hydrolysis.

$$CO(NH_2)_2 + 3H_2O \rightarrow 2NH_4^+ + 2OH^- + CO_2 \uparrow \quad (1)$$

$$CO_2 + H_2O \leftrightarrow HCO_3^- + H^+ \quad (2)$$

$$HCO_3^- + H_2O \leftrightarrow H^+ + CO_3^{2-} \quad (3)$$

$$\mathrm{Cell} - \mathrm{Ca}^{2+} + \mathrm{CO}_3^{2-} \rightarrow \mathrm{Cell} - \mathrm{CaCO}_3 \downarrow \quad (4)$$

Inorganic precipitation involves the precipitation of minerals in the absence of bacteria and relies heavily on the available atmospheric CO_2 (Equation 2). CO_2 helps in helps in the formation of bicarbonate (Equation 6), which later forms calcium carbonate (Equation 3).
Studies on abiotic systems identified calcite precipitation to be less effective in comparison to bacterial system [1]. The precipitation rate is especially high during the early stages of the precipitation process. Since bacteria take time to adjust to the environmental conditions, an abiotic approach could possibly be superior where pretreatment has not taken place.
This paper presents a study that investigates the carbonate precipitating bacterial stimulation from a soil sampled from Formby, UK and the main objective is to provide a preliminary analysis on the possibility of *insitu* application of biomineralisation towards ground improvement.

2. Materials and methods

2.1 Sample collection
The soil sample was collected from Formby national park, United Kingdom. The soil was collected at 30cm depth using an auger and transported to the lab inside zip lock bags. The soil was stored at 4°C prior to analysis.

2.2 Experimental system
Soil was added to a 50ml syringe column and liquid exchange was performed using a peristaltic pump with a 10 mm ID tube at 5 rpm (Figure 1). Cementation liquid consists of 500mM of Calcium chloride and Urea. Nutrient broth was added along with the calcium chloride and urea. The liquid exchange was performed every 2 days. The experiment was conducted at 15°C. The experiment was terminated when no liquid was collected during the liquid exchange. The experiments were performed in triplicates to study organic and inorganic precipitation. For inorganic precipitation, the soil was autoclaved to remove biological matter.

2.3 Carbonate titration and SEM imaging
Carbonate titration was performed using the protocol followed by [5] and SEM imaging was performed using the Zeiss EVO50 scanning electron microscope at the University of Wolverhampton.

IOP Conf. Series: Earth and Environmental Science **68** (2017) 012010 doi:10.1088/1755-1315/68/1/012010

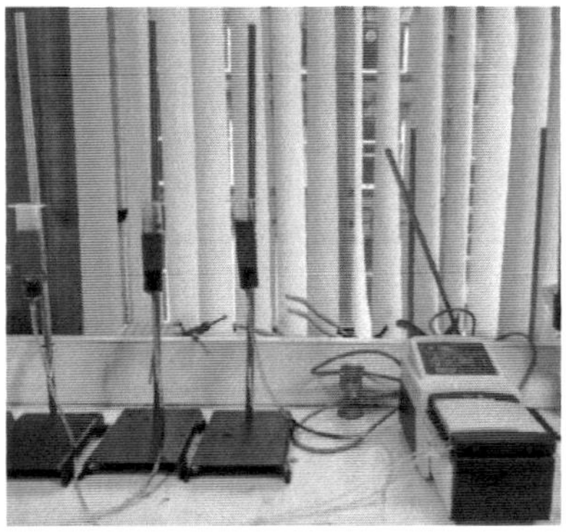

Figure 1. Experimental setup for the precipitation process.

3. Results and discussion

The experiments for organic and inorganic precipitation was conducted at 15°C. A temperature of 20°C is normally required to obtain an efficient calcite precipitation [6], [7]. However, the experiments conducted for this study were designed to mimic real _in situ_ conditions, thus the lower temperature of 15°C was maintained.

SEM images of particles after mineralization were used to identify the differing precipitation patters between the organic and inorganic samples (Figure 2 and 3). The organic precipitation produced snowflake like crystal formations on the top of the soil particles, such patterns being absent in the inorganic samples. The cementation bonds observed between soil particles due to organic precipitation are denser in comparison to the inorganic precipitation.

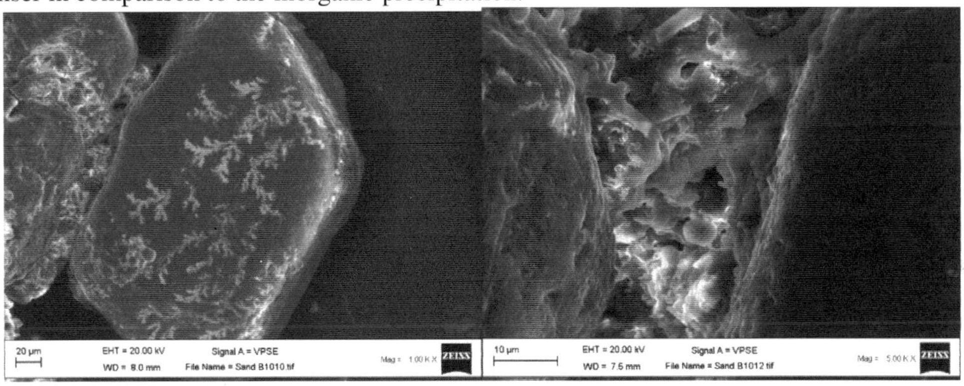

Figure 2. SEM images showing the cementation bonds through organic precipitation process.

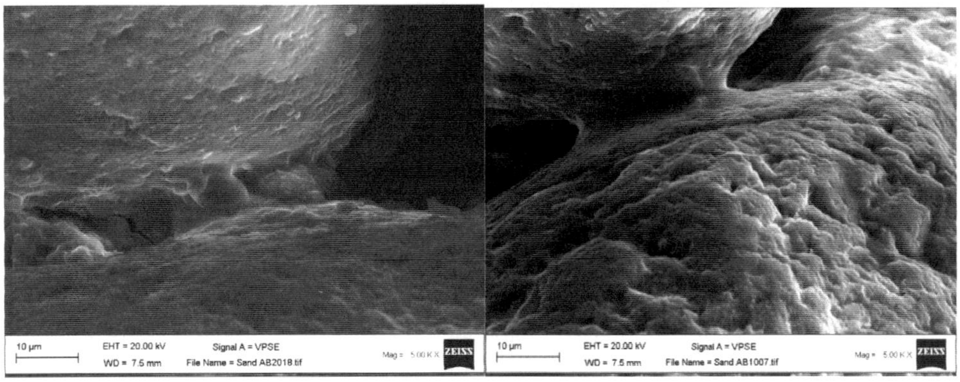

Figure 3. SEM images showing the cementation bonds through inorganic precipitation process.

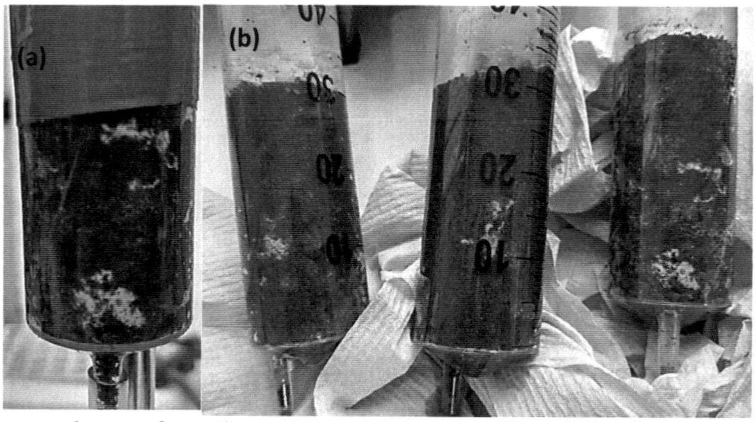

Figure 4. Calcium carbonate formations observed between soil particles during organic precipitation experiment (a) and observed in all three columns after the termination of experiments (b).

Before the drying of soil samples for carbonate titration, visible calcium carbonate formations were visible in the organic precipitation experiments. This was not observed for the inorganic precipitation experiments.

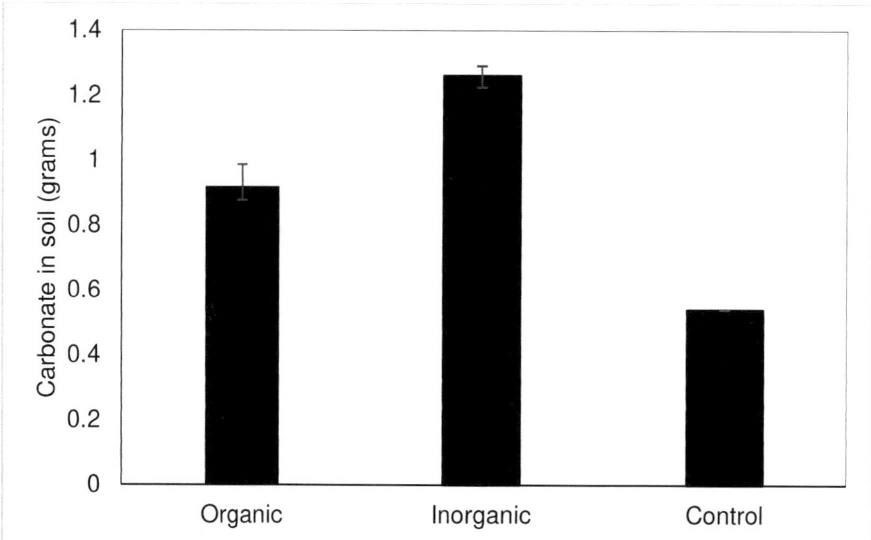

Figure 5. Carbonate precipitation obtained from organic, inorganic and control soil experiments. All experiments were performed in triplicates.

Carbonate titration identified that inorganic precipitation process precipitated 1.37 times more carbonate than the organic precipitation (Figure 5). The data comparison between the organic precipitation and control sample reveals that carbonate precipitating bacteria exist in the soil. The superior precipitation of carbonate through inorganic process is probably due to the additional time required for the organic process, since the bacteria take time to adjust to the environment.

4. Improvements that need to be made for future soil experiments

The preliminary data presented in this paper suggests that carbonate precipitating bacteria exists in soils and they could be stimulated for calcite precipitation. However, in this case, the organic precipitation was inferior in comparison to the inorganic precipitation, although both were superior to the control. Applied indigenous biomineralisation is still a new technique, a few limitations need to be addressed prior to implementation in the field;

MICP is a microbial processes which highly depends on temperature, pH, calcium concentration, DIC and the presence of nucleation sites [8]. This makes it a complex and time consuming process compared to the chemical process. Bacteria that's present in the soil can produce organic acids to counterbalance the pH increase associated with ammonia diffusion which results in the delay of crystallization. MICP has to be optimized for time effectiveness before it's used for *insitu* applications. Since the data shown in this study indicates that temperatures as low as 15°C could make organic precipitation less efficient in comparison to organic precipitation. Multiple parameter studies need to be conducted to optimize the efficiency and crystal productivity for organic precipitation.

5. Acknowledgments

This work was supported by grant no. PGRS-12-02-06 and RDF-13-01-06 awarded by XJTLU. I would like to thank the support of Dr Keith Jones for the electron imaging of the samples.

6. References

[1] Stocks-Fischer, S., J.K. Galinat, and S.S. Bang, *Microbiological precipitation of CaCO3.* Soil Biology and Biochemistry, 1999. **31**: p. 1563-1571.

[2] Bazylinski, D.A., et al., *Controlled Biomineralization of Magnetite (Fe3O4) and Greigite (Fe3S4) in a Magnetotactic Bacterium.* Applied and Environmental Microbiology, 1995. **61**(9): p. 3232–3239.

[3] Dhami, N.K., M.S. Reddy, and A. Mukherjee, *Synergistic Role of Bacterial Urease and* Carbonic *Anhydrase in Carbonate Mineralization.* Applied Biochemistry and Biotechnology, 2014. **172**: p. 2552-2561.

[4] Okwadha, G.D. and J. Li, *Optimum conditions for microbial carbonate precipitation.* Chemosphere, 2010. **81**(9): p. 1143-1148.

[5] Maulood, P.M., et al., *Comparison between Calcimetric and Titrimetric Methods for Calcium Carbonate Determination.* Open Journal of Soil Science, 2012. **2**: p. 263-268.

[6] Zamarreño, D.V., E. May, and R. Inkpen, *Influence of Environmental Temperature on Biocalcification by Non-sporing Freshwater Bacteria.* Geomicrobiology Journal, 2009. **26**(4): p. 298-309.

[7] Helmi, F.M., et al., *Calcium carbonate precipitation induced by ureolytic bacteria Bacillus licheniformis.* Ecological Engineering, 2016. **90**: p. 367-371.

[8] Ivanov, V. and J. Chu, *Applications of microorganisms to geotechnical engineering for bioclogging and biocementation of soil in situ.* Reviews in Environmental Science and Bio/Technology, 2008. **7**(2): p. 139-153.

IOP Conf. Series: Earth and Environmental Science **68** (2017) 012011 doi:10.1088/1755-1315/68/1/012011

Estimation of above ground biomass in boreal forest using ground-based Lidar

L Taheriazad[1], H Moghadas[2] and A Sanchez-Azofeifa[1]

[1] Earth and Atmospheric Sciences Department, University of Alberta, Canada
[2] Department of Mechanical Engineering, University of Alberta, Canada

Abstract. Assessing above ground biomass of forest is important for carbon storage monitoring in boreal forest. In this study, a new model is developed to estimate the above ground biomass using ground based Lidar data. 21 trees were measured and scanned across the plot area study in boreal forests of Alberta, Canada. The study area was scanned in the summer season 2014 to quantify the green biomass. The average of total crown biomass and green biomass in this study was 377 kg (standard deviation, S.D. = 243 kg) and 6.42 kg (S.D. = 2.69 m), respectively.

1. Introduction

Forest ecosystems deliver many viable goods and services, invaluble for human life [1], [2]. Forests provide ecological services such as regulation of water regime, preservation of soil quality, prevention of soil erosion, regulation of climate, protection of habitats for other species and conservation of biodiversity [3]. Forests are also the basis of industries such as timber, processed wood and paper, rubber, and fruits. They also help ecotourism by offering scenic and landscape services [3].

The proper assessment of forest benefits requires accurate calculation of forest structure that is described by parameters such as total leaf area and biomass. The amount of total leaf area defines functions such as radiation absorption, precipitation interception, photosynthesis [4] and the exchange processes between the atmosphere and the land surface [5]. Estimation of above ground biomass (AGB) at local, provincial and worldwide scales is basic for assessing global carbon storage and evaluating ecosystem reaction to environmental change and anthropogenic interferences [6], [7].

The study of the interrelation of biomass and leaf area is a strong tool for development of forest growth model [8], [9]. Due to the quantity of vegetation material in the forests retrieving the detail from structural elements by direct measurements is impractical. Remote sensing technologies such as LiDAR enable automatic collection of a detailed database on the forest structure and calculation of parameters such as leaf area and biomass [10]-[13]. Small-footprint discrete return ground based Lidar systems provide a high precision since data at individual tree level can be described [14], [15].This paper studies the interrelation between biomass and leaf area in Canadian boreal forest using ground based LiDAR.

2. Study area

The study area in this project is a 50 m × 50 m plot located in northwestern Alberta, Canada (latitude: 56° 44' 24" N; longitude: 118° 20' 24" W; altitude: 871 m) at the Peace River Environmental Monitoring Super Site (PR-EMSS) which consist of an old-growth stand of Trembling Aspen (*Populus tremuloides*) with a broad leaf deciduous canopy. The supersite is situated in the joint industry-research forestry region for Ecosystem Management Emulating Natural Disturbance (EMEND) for large-scale variable preservation boreal forest harvest experimentation. Figure1 shows the location map. At the study area, we measured 21 trees manually and scanned by ground Lidar. The mean value of tree heights by Lidar was 25.48 m (standard deviation, S.D. = 1.59 m) and mean value of diameter at breast height (DBH) was 38.6 cm (S.D. = 7.31 cm).

3. Material and methods

In this study, the point cloud data were collected from a high-elevation aspen stand using the terrestrial laser scanner Leica Scan Station C10. The plot scanning is shown in figure 2a. To avoid the shadowing effect and collect a complete point cloud, the site was scanned from five locations with 4.5 mm sampling spacing at 50 m. Meanwhile, five targets were set as reference points to align the five scans. The targets` locations were made observable in most of the scans. The experiment layout for locations is shown in figure 2b. The black signs represent the scan locations of Leica device. The characteristics of C10 are summarized in Table 1. Scanner locations are indicated alphabetically in the order of gaining. T1 to T5 shows the location of reference targets with a black dot. In this plot, north is shown. The distance between the points is marked in the figure. All five scans from the plot study were co- registered together using four standing targets. The collection, registration, and processing of the TLS point cloud data were performed in the Cyclone 8.1 software and MATLAB R2013b. The study area was scanned in summer 2014.

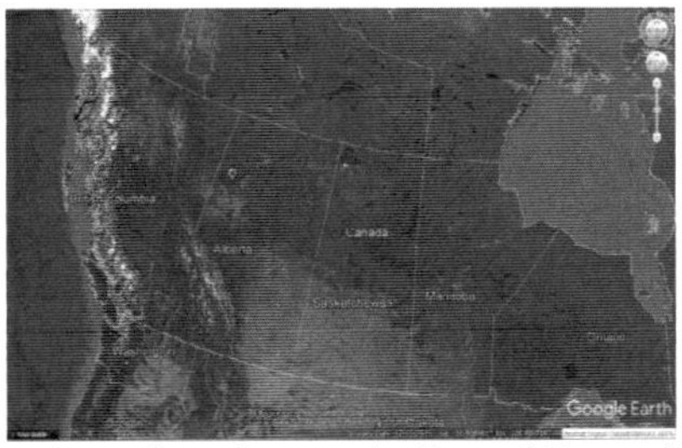

Figure 1. Location of study area in Boreal Forest NW Alberta.

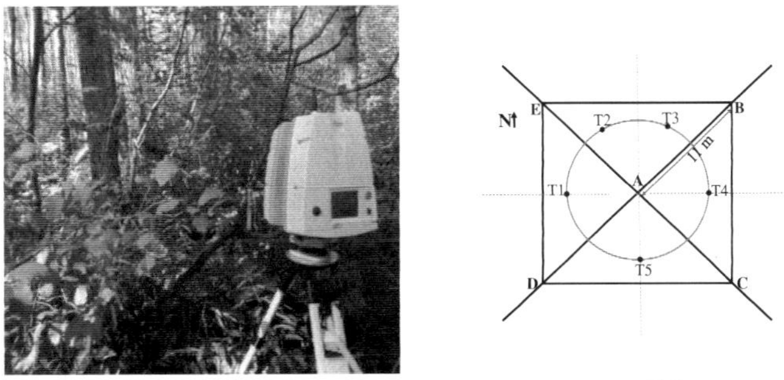

(a) (b)

Figure 2. Plot scanning with Leica C10 (a) and Schematic TLS field setup for plot study (b).

Table 1. Leica Scan Station C10 scanner settings for data acquisition.

Beam divergence	0.1mrad
Range	0.1-300m
Azimuth range	0–360∘
Max Zenith range	270
Acquisition time	6 min 45 s
data acquisition rate	50000 Samples/sec

In this study plant area (PA) of individual trees was estimated in a developed algorithm in MATLAB. Leaf area (LA) is defined as the amount of live leaf material in trees with the unit of m2 [16], [17]. The algorithm is applied on co-registered TLS return from five scanning stations based on contact frequency method.

IOP Conf. Series: Earth and Environmental Science **68** (2017) 012011 doi:10.1088/1755-1315/68/1/012011

In this study the tree height and DBH were also measured in the field manually, using laser Rangefinder Model and DBH measuring tape respectively (figure 3). Illustration of LiDAR measurements from one of the tree is shown in figure 4.

(a) (b)

Figure 3. Field measurement devices: a) laser Rangefinder and, b) DBH measuring tape.

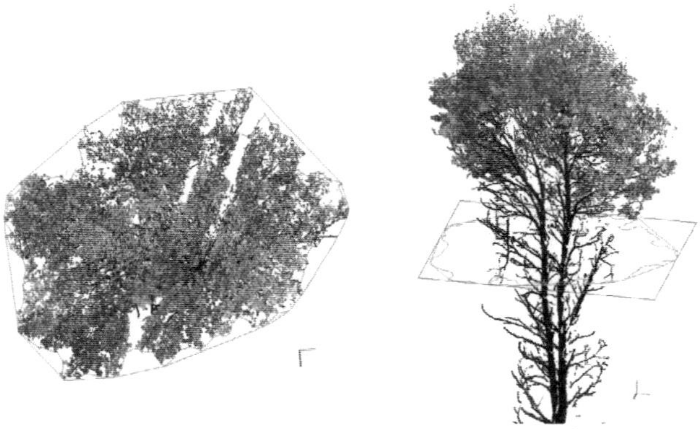

(a) (b)

Figure 4. LiDAR measurements: Projected plant area of individual tree top view (a), side view (b).

4. Results and discussion

The relationship between Lidar data and field data is shown in figure 5. There is a strong correlation ($r = 0.98$, $p<0.001$, $N=21$) between DBH calculated by Lidar and DBH calculated by field measurement while this correlation in tree height data is lower ($r = 0.80$).

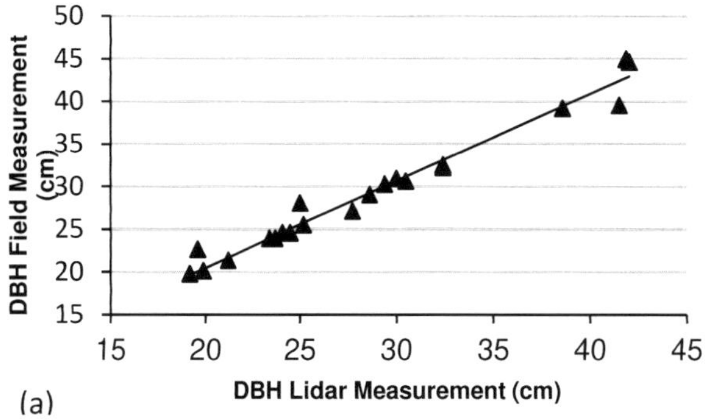

(a)

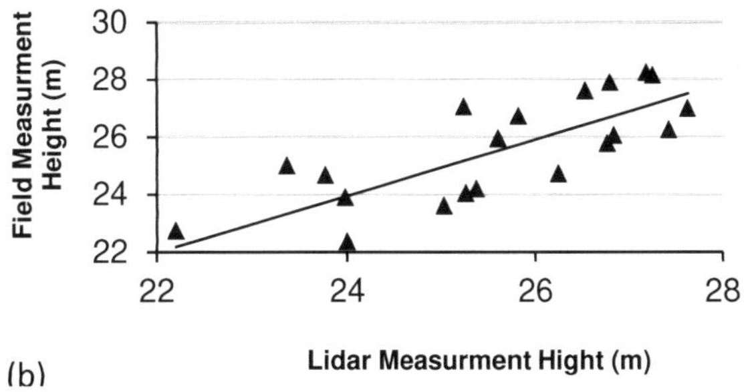

(b)

Figure 5. DBH (a) and Height (b) Relationship between Lidar measurement and Field measurement.

Total biomass calculated for individual trees in plot study is related nonlinearly to DBH and height in figure 6.

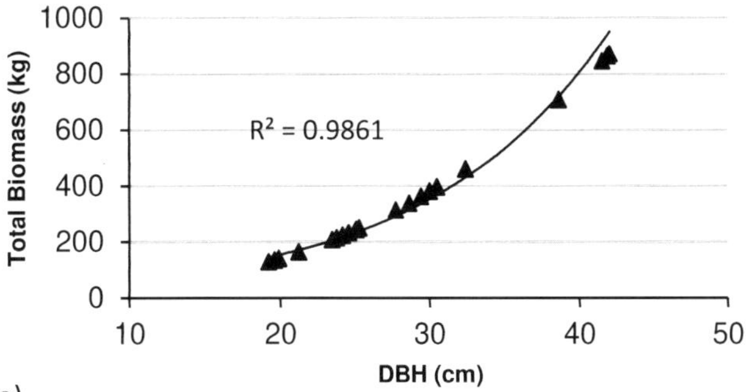

(a)

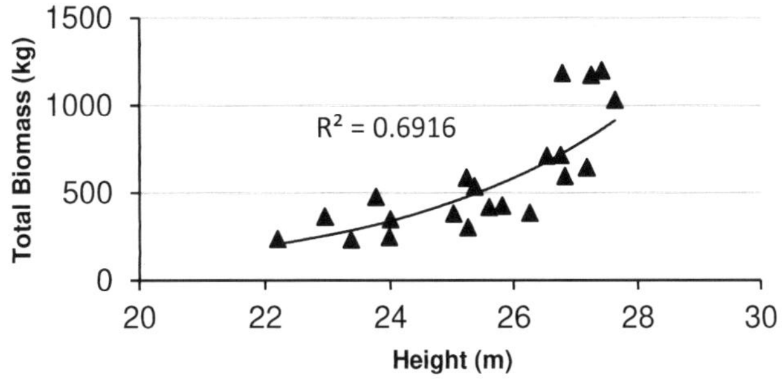

(b)

Figure 6. Relationship between total biomass and DBH (a) and height (b).

As figure 6a shows there is a strong correlation ($r = 0.98$, $p<0.001$, $N=21$) between total biomass and DBH. The correlation between total biomass and tree height is $r = 0.77$. Adding tree height as a predicting parameter to calculate biomass resulted in a slight decrease of the regression coefficients R^2. Crown biomass (branches and leaves) and leaf area are related to DBH and height in figure 7. Leaf area is calculated for each single tree in developed algorithm in MATLAB.

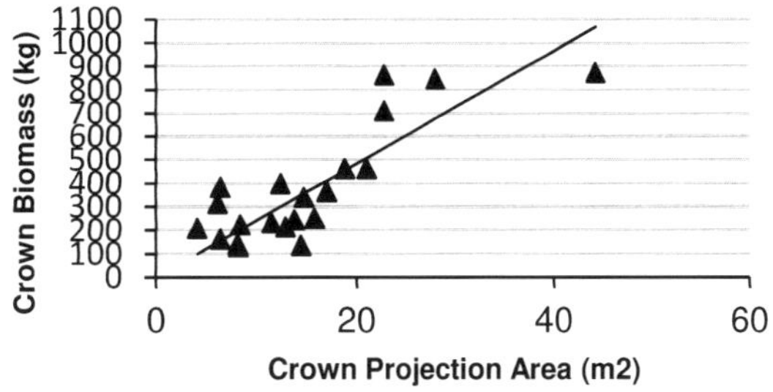

(a

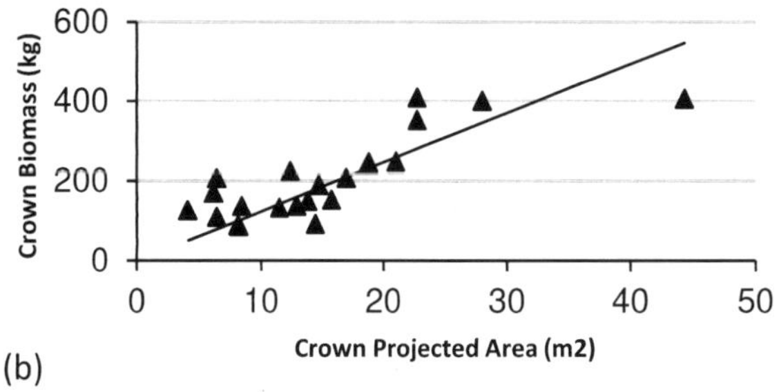

Figure 7. Relationship between crown biomass and plant area (a) using DBH model for calculating biomass and (b) using DBH and height model to calculate biomass.

The results in the figure 7 shows Leaf area and crown biomass are strongly linearly related ($r= 0.82$, $p<0.001$, $N=21$) for both models.

5. Conclusion

This paper presented a new method for estimation of above-ground biomass which was applied to a point cloud data from ground LiDAR. The terrestrial data was collected from a boreal forest at Peace River, Alberta, Canada during summer in 2014.

6. References

[1] Forest Europe 1993 Resolution H1, general guidelines for the sustainable forest management of forests in Europe, 16-17 June, Helsinki, Finland. Second Ministerial Conference on the Protection of Forests in Europe, Madrid, Spain.

[2] Millennium Ecosystem Assessment 2005 Ecosystems and human well-being: synthesis. Island Press, Washington, DC, USA.

[3] Lindberg K, Furze B, Staff M and Black R 1997 Ecotourism and other services derived from forests in the Asia- Pacific region: outlook to 2010.

[4] Parker GG 1995 Structure and microclimate of forest canopies. In: Lowman, M., Madkarni, N. (Eds.), Forest Canopies. Academic Press, California, USA, 73–106.

[5] Scholes R J, Frost P G H and Tian Y 2004 Canopy structure in savannas along a moisture gradient on Kalahari sands. *Global Change Biol.* 10, 292–302.

[6] Hese S, Lucht W, Schmullius C, Barnsley M, Dubayah R, Knorr D, Neumann K, Riedel T and Schröter K 2005 Global biomass mapping for an improved understanding of the CO2 balance – the Earth observation mission Carbon-3D. *Remote Sens. Environ.* 94, 94–104.

[7] Ni-Meister W, Lee S, Strahler A H, Woodcock C E, Schaaf C, Ranson J, Sun G and Blair J B 2010 Assessing general relationships between above-ground biomass and vegetation structure parameters for improved carbon estimate from vegetation lidar *J. Geophys. Res.* 115, G00E11, http://dx.doi.org/10.1029/2009JG000936.

[8] Jarvis P G and Leverenz J W 1983 Productivity of temperate, deciduous and evergreen forests. In. Physiological Plant Ecology. IV. Encyclopedia of Plant Physiology, New Series, Vol 2D (OL Lange, PS Nobel, CB Osmond, H Ziegler, eds), Springer-Verlag, New York, NY, USA, 233-280.

[9] Causton D R 1985 Biometrical, structural and physiological relationships among tree parts. In: Attributes of Trees as Crop Plants (MGR Cannell, JE Jackson, eds), *Inst Terrestrial Ecology*, Huntingdon, UK, 137-159.

[10] Nowak D J, Crane D E, Stevens J C, Hoehn R E, Walton J T and Bond J 2008 A Ground-based method of assessing urban forest structure and ecosystem services, *AUF* 34(6),347–358.

[11] Chen Q, Baldocchi D, Gong P and Kelly M 2006 Isolating individual trees in a savanna woodland using small footprint LIDAR data Photogram. Eng. Remote Sens. 72 (8), 923–932.

[12] Zhao F, Yang X, Schull M, Roman-Colon R, Yao T, Wang W, Zhang Q, Jupp D L B, Lovell J L, Culvenor D S, Newnham G J, Richardson A D, Ni- Meister W, Schaaf C L, Woodcoc C E and Strahler AH 2011 Measuring effective leaf area index, foliage profile, and stand height in

New England forest stands using Echidna® ground-based lidar *Remote Sens. Environ.* 155 (11), 2954–64.

[13] Yao T, Yang X, Zhao F, Wang Z, Zhang Q, Jupp D, Lovell J, Culvenor D, Newnham G, Ni-Meister W, Schaaf C, Woodcock C, Li X and Strahler A 2011 Measuring forest structure and biomass in New England forest stands using Echidna ground-based lidar *Remote Sens. Environ.* 115 (11), 2965–2974.

[14] Zhao F, Guo Q and Kelly M 2012 Allometric equation choice impacts lidar-based forest biomass estimates: A case study from the Sierra National Forest, CA. *Agr. Forest. Meteorol.* 165, 64– 72.

[15] Béland M, Widlowski J L, Fournier R A, Côté J F and Verstraete M M 2011 Estimating leaf area distribution in savanna trees from terrestrial LiDAR measurements *Agr. Forest. Meteorol.* 151, 1252–66.

[16] Watson D J 1947 Comparative physiological studies on the growth of field crops. I. Variation in net assimilation rate and leaf area between species and varieties, and within and between years *Ann. Bot.* 11, 41–76.

[17] Asner G P, Scurlock J M O and Hicke J A 2003 Global synthesis of leaf area index observations: implications for ecological and remote sensing studies *Global Ecol. Biogeogr.* 12 (3), 191–205.

IOP Conf. Series: Earth and Environmental Science **68** (2017) 012012 doi:10.1088/1755-1315/68/1/012012

Ecosystem Resilience to Drought and Temperature Anomalies in the Mekong River Basin

T Na-U-Dom[1,2], M García[2,3*] and X Mo[1,4]

[1]Key Laboratory of Water Cycle and Related Land Surface Processes, Institute of Geographic Sciences and Natural Resources Research, Chinese Academy of Sciences, Beijing 100101, China
[2]Sino–Danish college, University of Chinese Academy of Sciences, Beijing 100049
[3]Department of Environmental Engineering, Technical University of Denmark, 2800 Kgs. Lyngby, Denmark
[4]College of Resources and Environment, University of Chinese Academy of Sciences, Beijing 100049, China

E-mail: mgarc@env.dtu.dk

Abstract. Climate change is leading to an increasing in the frequency and intensity of extreme weather events, which significantly affect ecosystems stability. In this study, ecological stability metrics in response to wet/dry events and warm/cold events on vegetation greenness were assessed using an auto-regressive model of NDVI in the Mekong River basin (around 759,000 km^2) where large ecological and climatic gradients exist. Gridded temperature, and the Global Standard Precipitation Evaporation Index (SPEI) and antecedent NDVI were used as model predictors. The forest in north Laos was more resilient to the temperate and wet/dry anomalies events than other regions in the basin. Drought reduced green biomass in north Laos, northeast Thailand and Myanmar, but in these tropical climate regions' the vegetation biomass was also more responsive by higher temperatures. Vegetation in northeast Thailand, Cambodia and the Mekong delta were less sensitive to the temperature anomalies effect compared to other part of Mekong River basin. The map of resistance and resilience metrics can help to determine the most vulnerable regions to extreme events for policy makers.

1. Introduction

Climate change is considered as one of the major factors putting pressure on ecosystems worldwide. Climate models have found a connection between global warming and the occurrence of climatic extreme events, such as wet and dry anomalies (floods and droughts), in frequency, intensity and duration, which expected to significantly affect ecosystem [1]-[3], e.g. vegetation shift [4] and tree mortality [5]. Within this content, assessment of ecosystem stability to extreme events is crucial to qualify the sensitivity of ecosystems to environmental disturbance, but also to improve the understanding of the mechanism that govern vegetation response. The effect of extreme events on ecosystems after an event has been vanished has been referred to as "memory effect" [6]. Memory effects can be evaluated assessing the ecosystem response to climate-induced disturbance [7]. For this purpose, resilience and resistance metrics of ecosystem stability have been developed by analyzing

anomalies of an ecosystem property relative to their annual cycle. In this study we refer to ecosystem resilience as its the ability to recover after a disturbance, while its resistance expresses the response or sensitivity of ecosystem to an extreme events [8], [9].
To assess ecosystem resilience and resistance, it is possible to use different methods, such as lagged cross correlation analysis, canonical correspondence analysis or the persistence of NDVI trend [10]-[12]. Several studies have focused on the assessment of ecosystem stability, but there are not studies yet on the impacts of climatic anomalies on ecosystem in the Mekong River Basin (MRB). However, understanding the potential impacts of rainfall and temperature anomalies on vegetation is crucial in this region, home of approximately 70 million people and a highly biodiverse region.
The research aims to: (1) investigate both ecological resilience and resistance of ecosystems in the MRB to wet/dry anomalies and temperature anomalies using an ecological model based on remote sensing and climate data [7], [13] and (2) improve the understanding of ecosystem stability in the MRB. Analysis of vegetation responses to extreme events is important for ecological and environmental management programs, early warning and forecasting future climate-induced vegetation change in the MRB. Moreover, this research will help to identifying those regions in the MRB that are more sensitive to climate anomalies providing vulnerability maps, which can provide a guide for ecological and environmental management decisions in the MRB.

2. Study area

The MRB, which is shown in Figure 1, is originated in the Tibetan plateau in China. The basin covers six countries, is equivalent to 795,000 km^2, where the outlet run into the South China Sea (Figure 1a). The MRB climatic system is governed by the northeast and southwest monsoon [13]. The temperature and precipitation in the MRB are topographically uncertain place to place. The entire basin average temperature is around 24 ^{o}C and the range of precipitation ranges to 600 mm/year in the Tibetan plateau to more than 3,000 mm/year in Laos. The vegetation in the basin (Figure 1b) is unevenly distributed and associated with elevation, especially in the upper MRB, temperature, and precipitation system, especially in the lower MRB [14,15].

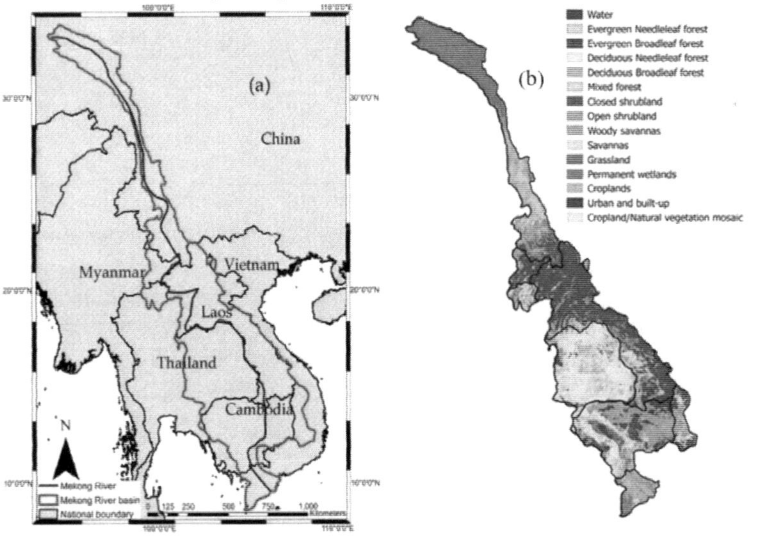

Figure 1. (a) The location, and (b) land cover, which retrieved from MODIS land cover type product (MCD12Q1) [16].

3. Materials and methods

The Normalized Difference Vegetation Index (NDVI) time series from January 1982 to December 2013 were downloaded from the Global Inventory Modeling and Mapping Studies (GIMMS) [17] for qualifying the response of ecosystem. The raw bimonthly NDVI image were filtered using Savitzky-Golay method and were resampled to monthly scale using the average of bimonthly NDVI image, then all images were reprojected as GCS-WGS1984. Monthly temperature data were downloaded from the

Climate Research Unit Time Series Version 3.23 (CRU-TS 3.23) [18] in the same time frame as NDVI dataset. Monthly standardize NDVI anomalies and monthly standardized temperature anomalies were derived by removing the seasonal variation of the NDVI and temperature time series. The drought effect was evaluated using the Standardized Precipitation – Evapotranspiration index (SPEI), is calculated from the average water balance, [19], [20] with the same temporal frame as NDVI time series data. In this study, 3-month SEPI was used to quantify extreme events effect to the ecosystem. Positive value represents wetter condition than average, while negative value shows dryer condition than average. The monthly temperature anomalies and 3-month SPEI data were resampling to GIMMS's resolution using a bilinear interpolation and were reprojected as GCS-WGS1984.

4. Model to assess ecosystem resilience and resistance

In this study, an auto-regressive model based on De Keersmaecker et al. [7] was applied to predict NDVI anomalies in each pixel based on climatic variables and antecedent NDVI to account for memory effects at every time step. The model parameters reflect the different sensitivity of vegetation to climatic and vegetation and thus allow to estimate ecosystem resilience and resistance to drought and temperature anomalies for each pixel. The model, shown as Equation 1, was applied every month in every pixel [7].

$$NANO_t = \beta\, NANO_{t-1} + \alpha\, SPEI_t + \varphi\, TANO_t + \varepsilon_t \qquad (1)$$

Where $NANO_t$ and $NANO_{t-1}$ are standardized NDVI anomalies at time t and $t-1$, respectively. $SPEI_t$ and $TANO_t$ are 3- month SPEI and standardized temperature anomalies at time t. ε_t represents the residual term of time t.β, α and φ are model parameters, which are related to ecosystem resilience and resistance metrics. Each model parameter can be related to a different ecosystem stability metric. φ represents vegetation response to instantaneous temperature change, e.g. a temperature resistance metric. Similarly, α is indicated a drought resistance metric. β is an indicator for the dependence of anomalies on the previous values, hence, it represents the ecosystem resilience metric. Where β is large (small), the vegetation has a strong (less) influence from the previous time step and the ecosystem recovers slowly (quickly) from perturbations. Based on De Keersmaecker et al. [7], this optimal model was selected based on the root mean square error (RMSE) of observed and modeled standardized NDVI anomalies. Only those pixels with RMSE < 0.9 were considered as a good fit and the model retained.

5. Results and discussion

The ecosystem resilience and resistance metrics from the model parameters are shown as Figure 2. Regarding the errors, the model showed good performance in the forest ecosystems, especially in the south China, Laos, and Cambodia; it was only poor fitted in the grassland ecosystem in the Tibetan plateau in China, and cropland in the northeast of Thailand and the Mekong delta in Vietnam. This was probably due to (1) low range of climate and vegetation variables variability, (2) poor quality of time series input data, or (3) vegetation response from the other disturbances, such as ecological management and irrigation [7]. (4) vegetation memory effects not accounted for.

IOP Conf. Series: Earth and Environmental Science **68** (2017) 012012 doi:10.1088/1755-1315/68/1/012012

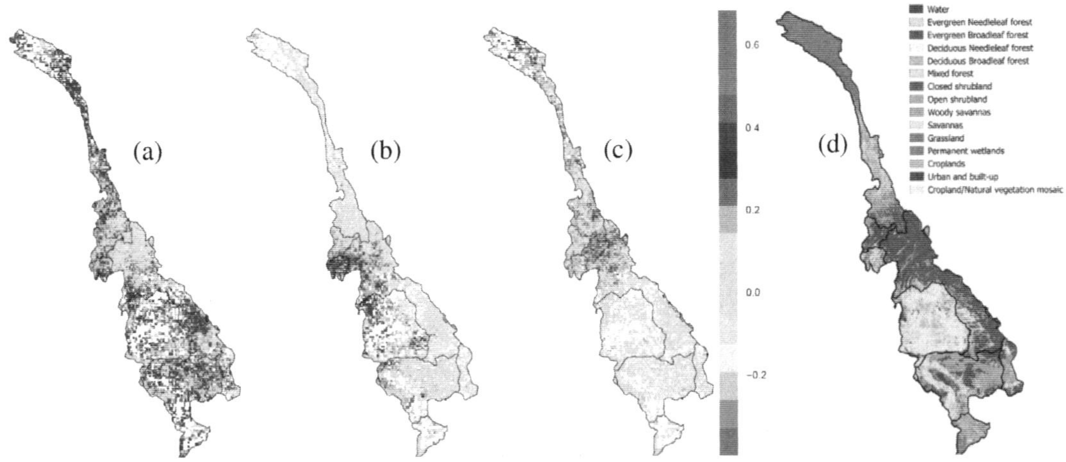

Figure 2. Model parameters which reflect (a) β, vegetation resilience, (b) α, indicated a drought resistance metric and (c) φ, temperature resistance metric. Pixels with RMSE > 0.9 are excluded. The colored bar represents the standardized coefficient scale. (d) Land cover in the Mekong River basin.

The vegetation resilience metric β (Figure 2a) is positive and large in the North Yunnan Province, China, in the south of Laos and Cambodia. It indicate that the ecosystem in those regions has low resilience or low recovery rates after climatic disturbances, needing more time to return to the equilibrium state. On the contrary, the evergreen forest in northern Laos showed high resilience, is indicated by values close to zero, with low recovery rates after disturbance more quickly return to the equilibrium state compared to the evergreen forest ecosystem in the south of Laos (Figure 2d).

From the vegetation drought resistance metric (α) (Figure 2b) it can be seen that vegetation in the upper part of basin has a negative response to drought events. Wetter (dryer) climate conditions decrease (increase) vegetation greenness, especially mixed forest biome, which widely spread in Yunnan province, evergreen forest biome in south Laos and Savanna in south Vietnam; however, the evergreen forest in north Laos, Myanmar and north Thailand shows positive value of vegetation drought resistance metric, e.g. low resistance to extreme climatic events, e.g., drought, and leading to decreasing biomass and vegetation greenness. The metric of vegetation resistance of temperature anomalies (Figure 2c) shows that vegetation in the north basin (cold climate) mostly shows a positive response with higher temperature anomalies, which increase vegetation greenness. Yet south Laos and northeastern Thailand and Cambodia shown a very large negative φ value that means the vegetation in these regions has low resistance (in the negative response) to temperature anomalies.

This modeling analysis is recommended to use when the ecosystem resilience and resistance will be monitored because the input data are used based on lag 1-month (or lag several month) of vegetation greenness data and several climate variables are represented updated climatic events, and could be taken into account which is useful to monitoring for preventing the ecosystem changing to un-equilibrium state. Other indicators, e.g. resistance and resilience metrics from cross-correlation analysis [10], can capture the time that when the vegetation will show the highest response to the climatic extreme event and this information could be incorporated to improve the modeling scheme.

6. Conclusion

This study applied a predictive model of NDVI anomalies based on multiple linear regression and auto-regression to assess the ecological stability due to wet/dry anomalies and temperature anomalies over the Mekong River Basin. The results showed good model fit for most of the pixels in all the basin (66 % of the basin) except the colder grasslands ecosystem in China and irrigated croplands in the northeast of Thailand and the Mekong delta in Vietnam. The ecosystem resilience and resistance metrics showed a clear pattern associated with the different land cover types in the MRB. The forest vegetation in the north Laos showed higher resilience to drought and temperature anomalies than other regions. Areas with low resilience and resistance, where drought affects the vegetation greenness included north Laos, the northeast Thailand and Myanmar. In this regions vegetation growth was linked to higher temperatures. Yet the vegetation in the northeast Thailand, Cambodia and the Delta showed less sensitive to the temperature anomalies effect. The provided maps showing more vulnerable regions to climatic extremes can guide management decisions in the MRB.

7. References

[1] International Development Research Centre 2015 *ScienceDaily* Available https://www.sciencedaily.com/releases/2015/10/151021104907.htm Accessed 2015 October 23.
[2] Intergovernmental panel on climate change 2012 *Summary for Policymakers. In Managing the Risks of Extreme Events and Disasters to Advance Climate Change Adaptation*. ed C B Field V Barros, T F Stocker, D Qin, D J Dokken, K L Ebi, M D Mastrandrea, K J Mach, G -K Plattner, S K Allen, M Tignor, and P M Midgley (Cambridge, UK and New York, NY, USA: Cambridge University Press)
[3] Walter F J 2012 Beyond productivity: Effects of extreme weather events on ecosystem processes and biotic interactions (Ph.D. Dissertation, University of Bayreuth, Germany). Available https://epub.uni-bayreuth.de/203/1/WalterDiss.pdf
[4] Lloret F , Escudero A, Iriondo J M, Martinez-Vilalta J and Valladares F 2012 *Glob. Change. Biol.* **18** . 797–805
[5] Lewis S L, Brando P M, Phillips O L, van der Heijden, G and Napstad, D 2010 *Science*. **331**. 554
[6] Wiegand T, Snyman H A, Kellner K and Paruelo J M 2004 *Ecosystems*. **8**. 243 – 258
[7] De Keersmaecker W, Lhermitte S, Tits L, Honnay O, Somers B and Coppin P 2015 *Global. Ecol. and Biogeogr*. **24**. 539–548
[8] Kang B, Lee S J, Kang D H and Kim Y O 2007 *J. Hydro-Environ. Res*. **1**. 118–125
[9] Harrison G W 1979 *Am Nat*. **113**. 659–669
[10] García M, Litago J, Palacios-Orueta A, Pínzon J E and Ustin S L 2010 *Appl. Veg. Sci*. **13**. 146–162
[11] Mitchell R J, Auld M H D, Le Duc M G, Robert M H 2000 *Perspect. Plant Ecol. Syst*. **3**. 142–160
[12] Harris A, Carr A S and Dash J 2014 *Int. J. Appl. Earth Obs. Geoinf.* **28**. 131-139
[13] Mekong River Commission 2005 Available: http://www.mrcmekong.org/topics/agriculture-and-irrigation/ Accessed 2015 July 3.
[14] Rundell P W 2009 *The Mekong: Biophysical Environment of an International River Basin* ed Campbell I C. (Amsterdam: Academic Press) chapter 7 pp 143–160
[15] Zhang B, Zhang L, Guo H, Leinenkugel P, Zhou Y, Li L and Shen Q 2014 *Int. J. Remote Sens*. **35**. 2835–2856
[16] Friedl M A, Sulla-Menashe D, Tan B, Schneider A, Ramankutty N, Sibley A and Huang X 2010 *Remote Sens. Environ*. **114**. 168–182
[17] Tucker C J, Pinzon J E, Brown M E, Slayback D A, Pak E W, Mahoney R, Vermote E F and El Saleous N 2005 *Int. J. Remote Sens*. **26**. 4485–4498
[18] Harris I, Jones T, Osborn T and Lister D H 2013 *Int. J. Climatol*. **34**. 623–642.
[19] Vicente-Serrano S M, Beguería S, López-Moreno J I, Angulo M and El Kenawy A 2010 *J. Hydrometeorol*. **11**. 1033–1043
[20] Beguería S, Vicente-Serrano S M, Reig F and Latorre B 2014 *Int J Climatol*. **34**.

IOP Conf. Series: Earth and Environmental Science **68** (2017) 012013 doi:10.1088/1755-1315/68/1/012013

Experimental study on removal of iron from potash feldspar

JW Kong, SQ Li, X Miao, JX Bai and CQ Yang

School of Metallurgical and Ecological Engineering, University of Science and Technology Beijing, Beijing 100083,China

Abstract. The potassium feldspar ore in a certain area of Chengde contains K_2O 9.82%, Na_2O 2.41%, which belongs to the high quality feldspar ore. It cannot be exploited and utilized because of the high iron content. An experimental research was carried out to remove iron from a feldspar ores in this area by the technique of high gradient magnetic separation respectively. The properties of potassium feldspar ore were analyzed, and the technology for iron removal from potash feldspar ore was expounded. The optimum parameters were determined. Finally, the content of Fe_2O_3 in the potash feldspar was reduced from the original 0.88% to 0.082%, significantly reduced the content of Fe_2O_3. The effect of iron removal was ideal, and the products can meet the requirements of high-grade daily porcelain on feldspar raw materials.

1. Introduction

Feldspar minerals are rich in potassium, sodium and other alkali metals. Feldspar minerals have low melting temperature (1100~1200℃) and long-time interval, widely used in glass, ceramics and other industrial sectors [1]. The impurities of natural potassium feldspar are quartz or iron oxide, Sodium oxide and calcium oxide followed [2]. The iron in feldspar ore is mainly composed of biotite, iron mineral, pyrite, iron containing silicate (such as amphibole, garnet, etc.). The use of iron ore grinding process will also be brought into part of the mechanical iron, and these iron impurities can be removed by magnetic separation, flotation, washing and other technical means [3]-[4]. In the process of magnetic separation, most of the strong magnetic minerals can be sorted out. But for some ores with high weathering degree, the iron ore is easy to slime in the grinding process, and high intensity magnetic separation is difficult to remove this part of iron minerals. The flotation process is better for clay and fine grained iron minerals. But for the weathered ore, iron removal effect is not very well. Generally speaking, the glass industry on the feldspar requirements are $K_2O+Na_2O>10.0\%$, $Fe_2O_3<0.3\%$, ceramic industry for higher iron requirements (general ceramic $Fe_2O_3<0.2\%$, high-grade ceramic $Fe_2O_3<0.1\%$) [5].

2. Experimental sections

2.1. Ore properties

The samples were collected from a region of Hebei, Chengde. After the crushing of massive feldspar, it is overall gray by visual observation. A comprehensive study of the results of X-ray diffraction analysis showed that the main minerals are feldspar, quartz, followed by a small amount of biotite and pyroxene, and iron minerals are mainly magnetite, limonite and other trace minerals are olivine. Results of multi-element analysis of samples are shown in table 1, and results of iron chemistry are shown in table 2.

Table 1 Analysis of multi element chemical composition of ore

Component/%	SiO_2	Al_2O_3	K_2O	Na_2O	Fe_2O_3	CaO	MgO	SO_3	SrO	TiO_2
Content/%	62.41	16.24	9.82	2.40	0.88	0.45	0.10	0.46	0.74	0.03

Table 2 Chemical phase analysis of iron in ores

Iron phase	Magnetite containing Fe	Limonite containing Fe	Silicate containing Fe	Total
Metal content /%	0.21	0.64	0.09	0.84
Distribution rate /%	21.61	64.54	13.85	100.00

As can be seen from table 1: the main ore elements are Si, Al, K, Na, and minor elements are Fe, Ti, Ca, Mg. The main components are SiO_2, Al_2O_3, K_2O, Na_2O, and as the content of Fe_2O_3 is as high as 0.88%, it cannot be directly used as raw material for ceramics and glass, so it should be excluded in the ore dressing. The content of K_2O in the ore is high, the content of K_2O and Na_2O is up to 12.23%, which can be used as the raw material of high quality ceramics, but the content of Fe_2O_3 in the mine is too high, resulting in low whiteness, which seriously affects its development and application.
As can be seen from table 2: iron mainly exists in the form of limonite, a small part of the existence of magnetite and other minerals. These iron minerals are either monomers or aggregates, highly dispersed in feldspar minerals, easy to remove; in the form of gangue minerals containing iron, such as biotite, fealties, hornblende, pyrite and so on.

2.2. Test flow

After the preliminary exploration test, we can determine the experimental process. Specific method: crushing raw ore, then grinding and screening, so that the feldspar ore 106μm through; Ore that is below the 106μm standard sieve has to remove the mud, taking the things that are on the top of sieve. The things that are on the top of sieve remove iron though superconducting high gradient magnetic separator, the optimum parameters were obtained by the contrast experiment. The iron removal test flow is shown in figure 1.

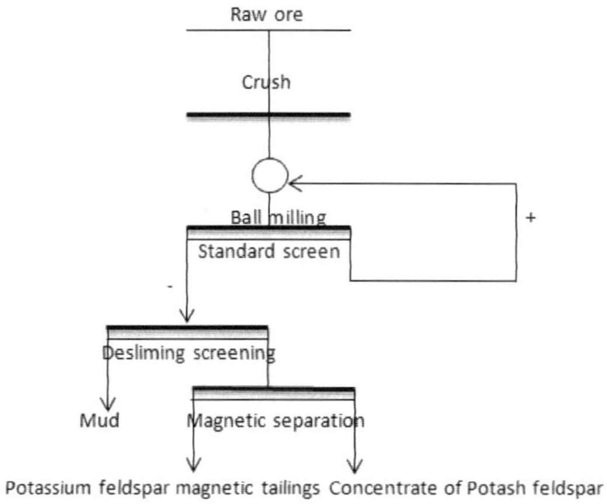

Figure 1 Iron removal test flow chart

IOP Conf. Series: Earth and Environmental Science **68** (2017) 012013 doi:10.1088/1755-1315/68/1/012013

3. Results and discussion

3.1. Test of grinding fineness

The test was carried out using a superconducting high gradient magnetic separator. Under the condition that the initial magnetic field intensity is 1.5T, the ore pulp flow rate is 500ml/min, the filling ratio of steel wool is 8%, and comparison results of different grinding fineness are shown in table 3.

Table 3 Test results of magnetic separation with different grinding fineness

Grinding fineness（-106μm）/%	Concentrate yield /%	Fe_2O_3 grade /%
50	32.32	0.42
60	43.55	0.39
70	56.12	0.27
80	64.63	0.22
90	68.25	0.32

It can be seen from the results of different grinding fineness magnetic separation tests: with the increase of grinding fineness, the productivity of concentrate increased gradually, and the content of Fe_2O_3 decreased firstly and then increased. When the content of the mineral that go though 106μm standard sieve is 80%, the content of Fe_2O_3 was the lowest. Considering the concentrate yield and Fe_2O_3 content, grinding fineness should be selected which the content of the mineral that go though 106μm standard sieve is 80%.

Table 4 Magnetic separation test results of magnetic induction intensity

Magnetic induction intensity /T	Concentrate yield /%	Fe_2O_3 grade /%
1.0	58.24	0.32
1.2	56.63	0.28
1.5	60.12	0.21
1.8	61.53	0.18
2.0	61.55	0.15
2.2	61.52	0.16
2.5	59.45	0.15

IOP Conf. Series: Earth and Environmental Science **68** (2017) 012013 doi:10.1088/1755-1315/68/1/012013

3.2. Test of magnetic induction intensity

Under the condition that the content of the particle size of -106μm is 80%, the ore pulp flow rate is 500ml/min, and the filling ratio of steel wool is 8%, the effect of different magnetic induction intensity on iron removal was investigated. Test results are shown in table 4.

It can be seen from the results of different magnetic intensity magnetic separation tests: with the increase of magnetic induction intensity, the content of Fe_2O_3 decreased firstly and then increased, and finally stabilized, the yield of mineral change little. Considering the concentrate yield and Fe_2O_3 content, and combined with the actual situation, the magnetic induction intensity should be selected 2.0T.

3.3. Test of ore pulp velocity

Under the condition that the magnetic induction intensity is 2.0T, the content of the particle size of -106μm is 80%, and the filling ratio of steel wool is 8%, the effect of removing iron from different slurry flow rate was investigated. Test results are shown in table 5.

Table 5 Test results of ore pulp velocity

Ore pulp velocity /(ml/min)	Concentrate yield /%	Fe_2O_3 grade /%
300	56.14	0.13
500	60.23	0.15
700	60.42	0.18
1000	62.24	0.36
1200	64.15	0.37

It can be seen from the test results of ore pulp velocity: with the increase of pulp flow rate, little change in concentrate yield, and the content of Fe_2O_3 increased gradually. Considering the concentrate yield and Fe_2O_3 content, as well as the actual situation, the ore pulp velocity should be 500 ml/min.

Table 6 Test results of filling rate of steel wool

Filling ratio of steel wool /%	Concentrate yield /%	Fe_2O_3 grade/%
8	62.36	0.15
10	61.52	0.14
12	61.34	0.11
15	57.35	0.16
18	50.45	0.23
20	45.34	0.28

3.4. Test of filling ratio of steel wool

Under the condition that the magnetic induction intensity is 2.0T, the content of the particle size of -106μm is 80%, and the ore pulp flow rate is 500ml/min, the effect of different steel wool filling rate on the iron removal index was investigated. Test results are shown in table 6.
It can be seen from the test results of filling rate of steel wool: with the increasing of the filling ratio of the separator, concentrate yield continued to decline, the content of Fe_2O_3 decreased firstly and then increased. Considering the concentrate yield and Fe_2O_3 content, combined with the actual situation, the filling ratio of the separator shall be 12%.

3.5 Multiple tests of superconducting high gradient magnetic separator
Under the experimental conditions of the optimum parameters of magnetic separation, the iron content of potash feldspar ore is 0.11%, which cannot reach the requirements of high-grade daily-use porcelain, so does magnetic separation test many times. Under the condition that the magnetic induction intensity is 2.0T, the content of the particle size of -106μm is 80%, the ore pulp flow rate is 500ml/min, and the filling ratio of steel wool is 8%, the results of multiple magnetic separation tests are shown in table 7.

Table 7 Test result of filling rate of steel wool

times of magnetic separation /times	Concentrate yield /%	Fe_2O_3 grade/%
Two times	61.35	0.094
Three times	62.23	0.082

It can be seen from the results of multiple magnetic separation tests: after two times, three times magnetic separation, the content of Fe_2O_3 in potash feldspar ore was obviously decreased, and the grade of concentrate is increased slightly. After three magnetic separation tests, the content of Fe_2O_3 reduced to 0.082%, which could meet the requirement of high grade domestic porcelain.

4. Conclusions

The results that an experimental study on removing iron from potash feldspar ore in a region of Hebei, Chengde, show:
(1) Though magnetic separation test, the optimum parameters of magnetic separation were determined: magnetic induction intensity 2.0T, the grinding fineness is which the content of the particle size of -106μm is 80%, and ore pulp velocity 500 ml/min, Steel wool filling rate 12%.
(2) The superconducting high gradient magnetic separator can effectively remove the iron of ore minerals. Raw ore is contained Fe_2O_3 0.88%, after one time superconducting high gradient magnetic separator magnetic separation can obtain feldspar concentrate which is contained $Fe_2O_3$0.11%, and concentrate yield is 61.34%. After two times, three times magnetic separation, it can obtain feldspar concentrate that is contained Fe_2O_3 0.082%, and concentrate yield was 62.23%.

5. References

[1] Q.M. Nie, L.Y. Li, X.X. Xu and GH Ai 2016. Experiment Study on Iron Removal of a Potash Feldspar Ore. *Non-Metallic Mines*. **39**(02), 8-10.
[2] Q. Cheng, R.C. Ren and C.X. Li 2013. Experimental Research on Purifying Low-grade Feldspar Ore from Liaoning by Mineral Processing Method. *Non-Metallic Mines*. (03), 41-42.
[3] X.J. Liu 2010. Experimental Study on the Iron Removal from the Tailing of a Potash Feldspar in Jiangxi Province. *Conservation and Utilization of Mineral Resources*. (03), 20-22.
[4] W.J. Li, T.B. Yue, L. Lv, F. Cao and L.H. Fang 2014. Removal of Iron Impurities from Potash Feldspar. *Mining engineering*. **34**(01), 44-46.
[5] H.C. Huang, T. Xiong and M.F. Xie 2016. Iron Removing Research and Application of SLon VPHGMS in Feldspar Ores. *Non-Metallic Mines*. (02), 72-74.

IOP Conf. Series: Earth and Environmental Science **68** (2017) 012014 doi:10.1088/1755-1315/68/1/012014

Preparation of the Iron Oxide Red from the Converter Dust by the Magnetic Separation and Roasting Process

Z J Guo, S Q Li, C Q Yang

School of Metallurgical and Ecological Engineering, University of Science and Technology Beijing, Beijing 100083, China

Abstract. Preparation of iron oxide red (α-Fe_2O_3) from the converter dust by the superconductivity high gradient magnetic separation (S-HGMS) and roasting process was investigated in the paper. The basic properties of the dust were studied by the X Ray Fluorescence, the chemical analysis and the X Ray Diffraction methods. The results showed that the raw dust mainly contained elements of Fe、O、Si、Ca, the iron content of the raw dust was 61.80%, and there were ferrous phases of Fe_3O_4, α-Fe_2O_3, $Fe_2(SiO_4)$ and $CaFe(Si_2O_6)$ in the raw dust. Under the optimum conditions of magnetic field intensity of 1.8T, the dispersion agent of 30mg/L and velocity of 500mL/min, the powders absorbed by the magnetic medium mainly contained Fe_3O_4 and α-Fe_2O_3, and the iron content of powders absorbed was up to 65.90%. The Fe^{2+} content of the powders absorbed under the optimum magnetic conditions dropped to 0.25% from 19.10% after roasting of fifty minutes, and the iron content of powders absorbed under the optimum magnetic conditions fell to 64% due to oxidation, and the Fe_3O_4 was removed. Finally the α-Fe_2O_3 content was up to 91.07% in the iron oxide red.

1. Introduction

In recent years, the converter dust mushroom with the rapid development of the iron and steel industry in china. Due to the high iron content, the dust is usually used as coolant in the converter by making the heated pressed block [1]-[2], and the dust also can be used as sintering burdening [3]. But the K, Na, Zn, S in the dust are bad for the ironmaking and steelmaking process or the steel quality. The high-iron-content dust can also be made into the high purity α-Fe_2O_3. The α-Fe_2O_3 has excellent light fastness, high temperature, alkali resistance and corrosion resistance. It can be made into the high temperature resistant coatings [4], the anticorrosion coating [5], strong acid resistance floor coating [6], ceramic colorant [7] and nano iron oxide material [8]. The dust contains magnetite and weakly magnetic hematite, and more weakly magnetic minerals. The hematite only is absorbed by high magnetic field. The experiment magnetic equipment in the study has the maximum magnetic intensity of 5T.

2. Experimental

2.1. The raw material

The chemical compositions and ferrous phases of the converter dust were studied by the XRF-1800 X Ray Fluorescence, the chemical analysis and the D8 ADVANCE Cu X Ray Diffraction. The results were showed as the table 1 and the figure 1.According to the figure 1, Fe is in form of Fe_3O_4, α-Fe_2O_3, $Fe_2(SiO_4)$, and $CaFe(Si_2O_6)$. Fe^{2+} is in the Fe_3O_4, $Fe_2(SiO_4)$, $CaFe(Si_2O_6)$. The magnetic susceptibilities are showed as the table 2.

Table 1. Main chemical compositions of the dust/%.

Element	TFe	Fe^{2+}	CaO	SiO_2
Content	61.80	19.10	4.42	1.41

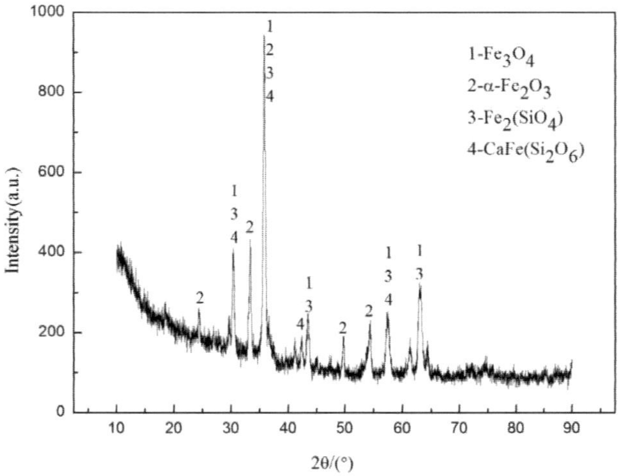

Figure 1. X-ray diffraction patterns of ferrous phases of the dust.

Table 2. The susceptibilities of minerals/$4\pi\times10^{-6}$SI.

Phase	Fe_3O_4	α-Fe_2O_3	$Fe_2(SiO_4)$	$CaFe(Si_2O_6)$
Susceptibility	100000~570000	50~4000	2~160	40~467

2.2. The process route

The magnetic intensity is the decisive fact, and the particle dispersion and velocity of flow are the secondary facts. The optimum conditions which Fe_3O_4 and α-Fe_2O_3 were absorbed mostly, and $Fe_2(SiO_4)$ and $CaFe(Si_2O_6)$ were not absorbed as far as possible by the magnetic medium was found in the experiment. The experiment was based on the other certain empirical facts, such as the size and concentration of particles, and the magnetic medium filling ratio. The powders absorbed were roasted in the temperature of 250°C, because Fe_3O_4 turned into α-Fe_2O_3 [9] under 250°C. The process route was showed as the figure 2.

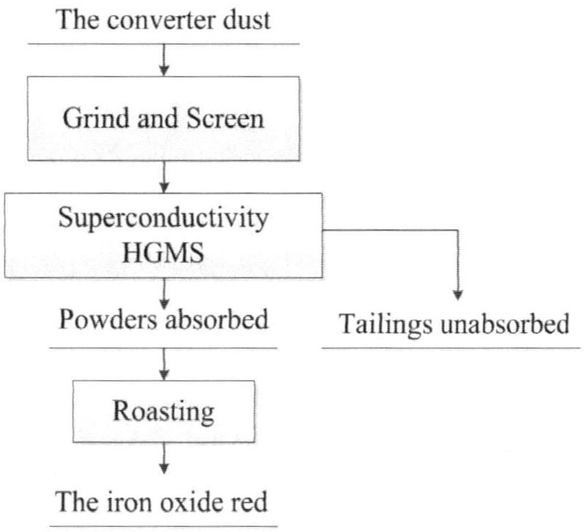

Figure 2. The process route.

2.3. Results and discussion

2.3.1. The influence of the magnetic field intensity on the magnetic separation

This section was under the conditions of velocity of 500mL/min, at -200 mesh and with 8g steel wool and without the dispersion agent. The result was showed as the figure 3.

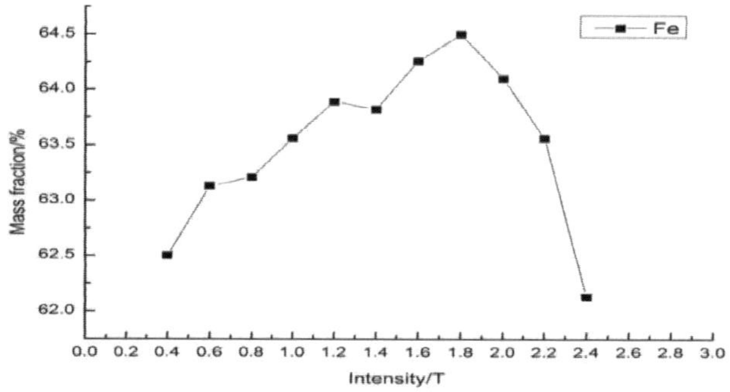

Figure 3. The influence of the magnetic field intensity on the magnetic separation.

The iron content of Fe_3O_4 is 72.41%, the iron content of α-$Fe_2O_3$is 70%, the iron content of $Fe_2(SiO_4)$ is 54.90%, the iron content of $CaFe(Si_2O_6)$ is 22.58%. According to the figure 3, in a certain range, the iron content of powders absorbed increased with the rise of the intensity, so the total content of Fe_3O_4 and α-Fe_2O_3 increased in this range. When the intensity reached 1.8T, the iron content reached the maximum of 64.50%, then the iron content declined with the rise of the intensity. The magnetic force of the particle was showed as the formula 2-1.

$$F_m = KVH\frac{dH}{dx} \qquad (2-1)$$

According to the formula 2-1, the magnetic susceptibility K, the particle volume V and the magnetic gradient dH/dx are certain, thus F_m is proportional to the magnetic intensity H. Fe_3O_4 is totally absorbed under the low H , and when the intensity is under 1.8T, the more higher the H is, the more α-Fe_2O_3 are adsorbed, when the intensity is out of 1.8T, the more higher the H is, the more particles which the H is less than α-Fe_2O_3 are adsorbed by steel wool, so the total content of $Fe_2(SiO_4)$ and $CaFe(Si_2O_6)$ increases.

2.3.2. The influence of the dispersion agent on the magnetic separation

This section was under the conditions of 1.8T and velocity of 500mL/min, at -200 mesh and with 8g steel wool. The result was showed as the figure 4.

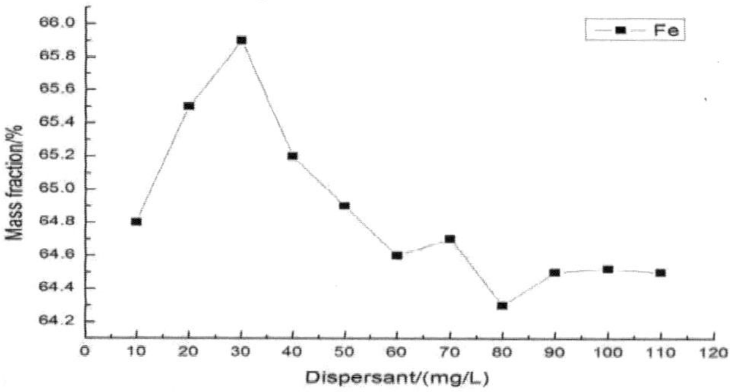

Figure 4. The influence of the dispersion agent on the magnetic separation.

According to the figure 4, in a certain range, the iron content of powders absorbed increased with the increasing of the dispersion agent. And when the dispersion agent concentration reached 30mg/L, the iron content reached the maximum of 65.90%, then the iron content declined with the rise of the dispersion agent.

$Na_6(PO_3)_6$ ionizes out the PO_3^{-1}, the PO_3^{-1} is absorbed by the particles surface, which is called the surface ion. Ions with the opposite charge in the medium are called the antiparticle. Some antiparticles are absorbed by the static electricity, which forms the close layer with negative charge. The antiparticles around the particles form the diffused layer. So the particles can repel each other by the electric double layer. But if the agent exceeds 30mg/L, more antiparticles will be absorbed by the close layer, the potential declines until the electric double layer disappears.

2.3.3 .The influence of velocity of flow on the magnetic separation

This section was under the conditions of 1.8T and dispersion agent of 30mg/L, at -200 mesh and with 8g steel wool .The result was showed as the figure 5.

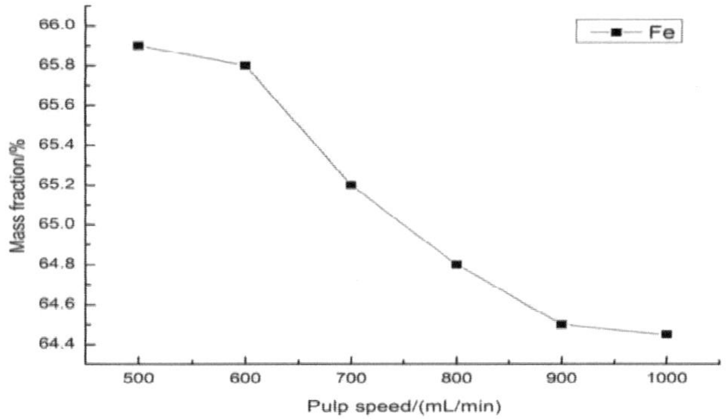

Figure 5. The influence of velocity of flow on the magnetic separation.

According to the figure 5, the iron content declined with the rise of velocity of flow. When velocity of flow reached 500mL/min, the iron content reached the maximum of 65.9%.The phases were mainly Fe_3O_4 and α-Fe_2O_3. The fluid drag force of the particle is as the formula 2-2.

$$F_n = 12\pi uvb \qquad (2-2)$$

According to the formula 2-2, the kinematic viscosity u, the particle relative speed v and the particles radius b is certain. The F_n is proportional to v, so the iron content declines with the increasing of velocity of flow.

2.3.4. The effect of roasting time on Fe^{2+} content of powders absorbed under the optimum magnetic conditions

The powders absorbed under the optimum magnetic conditions were roasted in the air under 250°C with the different time. Fe_3O_4 turned into α-Fe_2O_3.The result was showed as the figure 6.

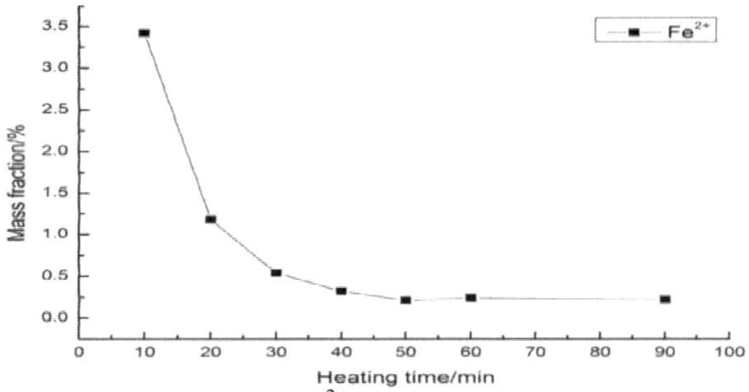

Figure 6. The effect of roasting time on Fe^{2+} content of powders absorbed under the optimum magnetic conditions.

According to the figure 6, the Fe^{2+} content declined with the increasing of the roasting time. The Fe^{2+} content reached 0.25% after the roasting of fifty minutes. Then the Fe^{2+} content was almost unchanged after more roasting time, so Fe_3O_4 was removed. The iron content of powders absorbed under the optimum magnetic conditions fell to 64% due to oxidation.The particle morphology of iron oxide red was showed as the figure 7.

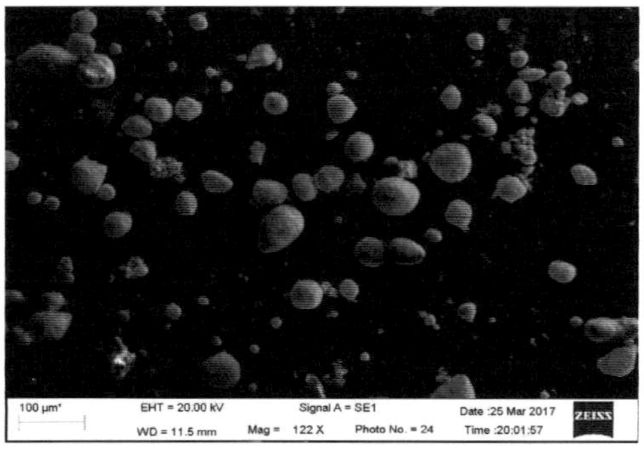

Figure 7. The particle morphology of iron oxide red.

3. Conclusions

The converter dust mainly contains magnetite, hematite, iron olivine, hedenbergite phase. Under the optimum conditions of magnetic field intensity of 1.8T, the dispersion agent of 30mg/L and velocity of 500mL/min, the powders absorbed by the magnetic medium mainly contained Fe_3O_4 and α-Fe_2O_3, and the iron content of powders absorbed was up to 65.90%. The Fe^{2+} content of the powders absorbed under the optimum magnetic conditions dropped to 0.25% from 19.10% after roasting of fifty minutes, and the Fe_3O_4 was removed. Finally the α-Fe_2O_3 content was up to 91.07% in the iron oxide red.

4. References

[1] Hong J G and Zheng H W 2015 Technical research and application of dry cold pressing block of the converter LT dust. Proceedings of the 2015 Annual Conference on refractories (the Thirteenth National Symposium on unshaped refractories and the Academic Symposium on refractories of the 2015) Thesis collection (3).

[2] Wan X F, Cao D, Gao X Z 2014 Research and application of the metallurgical dust sludge in converter. China Metallurgy, **24(11)** 35-41.

[3] Meng X F 2015 Comprehensive application of dust in sintering process. Heilongjiang Metallurgy, **36(5)** 59-61.

[4] Li Y N 2013 Development and evaluation of fluoro silicone resin coatings. (Jinan, University of Jinan).

[5] Chen Z and Yang Y P 2016 Study on the permeable closed waterborne epoxy anticorrosive primer. Jiangsu Construction, (Suppl) 79-82.

[6] He S P, Gao Q, Xu X T 2011 Method for preparing iron oxide red and its application in coatings. Chemical production and technology, **18(1)** 29-32.

[7] Qin W 2011 Application of iron oxide red pigment in ceramic industry. Foshan Ceramics, **(7)** 9-12.

[8] He M L, Li W, Su X 2015 Research progress of nano magnetic materials in water treatment. New chemical materials, **43(11)** 28-29.

[9] Liu X M, Jiang J Z, Mao X F 2010 The analysis of several types and characteristics of the magnetic hematite. Chinese Science, **40(5)** 592-601.

The effect of hydrochemical and other environmental variables on zooplankton communities in artificially connected ponds Highgate, London

Wen Liu[1], Zhirui Peng[2]

[1]Engineering in University of Cambridge, UK.
[2]Shenzhen Jianwen Foreign Language School, China

Abstract. Hydrochemical and other environmental variables can have impacts on zooplankton communities in artificially connected ponds. We examined five ponds environmental parameters by using lab equipment to identify and quality the zooplankton subsamples. From Shannon diversity figure and RDA analysis, indeed, environmental variables can affect the zooplankton communities in the ponds, and zooplankton communities varied in the different range in each pond.

1. Introduction

1.1. Natural and artificial aquatic bodies
Ponds, either natural or man-made, are often looked as a significant aquatic eco-system for farmland irrigation, flood protection, water supply, and fish production [1]. In the global scale, natural and artificial ponds still represent major percentage of the surface area of the standing freshwater resource, not all of them are in uniform distribution. Ponds and lakes, as small water reservoirs, are not expected to be rich in terms of biodiversity, because they tend to be physically and biologically unstable [2]. Mutual effects between physical and chemical profiles to water bio species are strongly connected with each other and bidirectional. Ecologically, many aquatic species and zooplanktons are important in defending surface water quality in many agricultural landscapes. Trophic state has been reported as a dynamic in biological abundance in aquatic area, such as nitrogen and phosphorus. However, exceeding nutrients will cause a serious environment issue, such as eutrophication; lack of fundamental nutrients may destroy the food chain in freshwater bodies [3].

1.2. Zooplankton

1.2.1. Basis information of zooplankton
Zooplankton is the common name given to many small species of animals found in fresh and marine waters throughout the world. The word zooplankton, derived from Greek, means "wandering animals." [4]. Planktons are organisms drifting in oceans, and bodies of fresh water. Zooplankton is an important ecological parameter in water monitoring and management schemes. Zooplankton species monitoring has been used widely in developed area research.

1.2.2. General factors which can influence the freshwater zooplankton community
Zooplankton community structure can be affected by biotic and abiotic factors within the freshwater zones. An abundance of zooplankton is related to many factors, such as water hydrochemistry, season, lake morphology, presence of macrophytes, predators etc., and related to the productivity of a lake, for example, lake trophic condition [5]. Global warming can influence the zooplankton directly. Most species of phytoplankton and

zooplankton can endure temperatures more than those now being forecast. Relatively subtle shifts in species dominance are brought form changes in the frequency and intensity of wind-induced mixing [6].

A study suggested that environment indexes, such as altitude, agricultural development, and geological morphology, in both local and regional can produce bio-patterns of zooplankton species composition and diversity in reservoirs [7].

Both natural profiles and anthropogenic changes in lake and watersheds can potentially influence zooplankton community structure [8]-[10].

Hydro-chemical materials in pond and lakes can literally affect freshwater zooplankton community in uncertain extents, such as total phosphorus, dissolved oxygen, soluble reaction phosphorus, and Nitrite Nitrogen. Some zooplankton studies focused on the impacts of ecosystem connectivity [11]. Examined zooplankton communities in ponds with and without artificial connections, and found that unconnected ponds were significantly correlated to environmental and distance factors.

1.3. Aims

This project aimed to access the zooplankton communities in five artificially connected ponds

The objectives were: (1) to provide the description of zooplankton communities, (2) to explore to discuss the relationships between the environment variables of the ponds and zooplankton communities

The main hypothesis: (1) total phosphorus can significantly influence the zooplankton communities; (2) Dissolved Oxygen is the subordinate factor of the zooplankton communities.

2. Method

2.1. Study sites

Highgate ponds chain was initially dug in the 17th and 18th centuries and currently is located in the eastern of Hampstead Heath Park, London. Five freshwater ponds constitute the Highgate chain; they are Stork pond, Bird Sanctuary pond, Model Boating pond, Highgate Men's Bathing Pond, and Highgate No.1 pond. The elevation gradient of Highgate Chain is from 87.97m to 65.5m, and all of the ponds are artificially connected by one or two water channels (which called spillway in Highgate Chain). Thus, the aquatic chain has high intra-communication in water chemistry and biology.

Stock pond (area, 4,401 m^2; elevation, 87.97 m) is located in the north part of Highgate Chain and completely surrounded by deciduous vegetation. Bird Sanctuary pond (area, 7,694 m^2; elevation, 78.20 m) is surrounded by overhung deciduous. The pond littoral zone was dominated by emergent macrophytes with no record of submerged vegetation. Model Boating pond (area, 16,280m^2; elevation, 76.01 m) is an open and public accessible pond. Little bankside vegetation could be found in the area because of the surrounding reinforced concrete. Men's Bathing Pond (area, 18,250 m^2; elevation, 70.8m) is the deepest and largest pond in this pond chain. The majority of the pond is surrounded by bankside trees in east and west sides. Highgate NO.1 pond (area, 13,660 m^2; elevation, 65.5m) is located in the south of the Highgate Chain. Dog is only allowed in this pond and can cause physical disturbance to an uncertain extent.

2.2. Sampling Methodology

2.2.1. Water sampling

All sampling bottles should be washed by 2% hydrochloric acid, distilled water and deionized water in order. In each pond, two 250 ml un-filtered water samples (open water and littoral zone) were collected for Total Phosphorus (TP) analysis using bottle samples. And for Nitrate Nitrogen and Soluble Reactive Phosphorus analysis, two 250ml filtered water samples were collected for each site.

2.2.2. Zooplankton sampling

Vertical core tube and a 53-ηm mesh Wisconsin bucket zooplankton net have been used to collect zooplankton. Zooplankton samples were taken from open water and littoral zone parts of each pond.

2.3. Laboratory works

2.3.1. Water chemistry and chlorophyll α analysis

Chlorophyll α concentration can be calculated as the following equation:

$$chloropgyll\alpha = \frac{11.0(A663)\ (v)}{V\ (\mathrm{d})}$$

Where: A663=absorbance at 663
v=volume (ml) of 90% acetone used for extraction (in this case 10ml)
V=volume of water filtered (in liters)
d=path length (cm) of the spectrophotometer cell (in this case 1cm)

2.3.2. Zooplankton identification and count
Using counting cell (1ml with 20×50 micro squares) and Leico microscope is the way to identify and quantify the zooplankton subsamples. Open water and littoral zone zooplankton communities were counted and recorded separately.

2.4. Cluster analysis
Diversity in each pond was used to compare the species diversities between connected ponds. To reveal the relationships between zooplankton communities and environmental parameters, canonical ordination was used in analysis.
Fish predation level, dissolved oxygen, Chlorophyll α, total phosphorus, and nitrogen were picked as environmental variables in RDA analysis systematically.

3. Result

3.1. Limnological characteristics

3.1.1. Conductivity in the ponds
According to the data taken (Tab.1), the conductivity in Bird Sanctuary Pond was the highest among Highgate Ponds Chain (460.5). The conductivity of Men's Bathing Pond (385.71) was the lowest.

3.1.2. Chlorophyll α in the ponds
Based on the presented data (Tab.1), Highgate NO.1 Pond which is located in the south of the Highgate Chain had the highest Chlo α (172.85μg/L). The top pond named Stock Pond had lowest Chlo α in the result.

Table 1. The basic information of the highgate ponds

Pond code	Stock	Bird	Boating	Men's	No.1
Elevation (m)	87.97	78.2	76.01	70.8	65.5
Area (m^{-2})	4401	7694	16780	18250	13600
Volume(m^{-3})	3400	7500	32900	42900	22300
Conductivity	447.33	460.5	421.62	385.71	393.57
Chlo a ($\mu g \cdot L^{-1}$)	9.48	17.26	19.9	65.86	172.85

3.2. Zooplankton in each pond

3.2.1. Zooplankton species
Highgate Ponds Chain was the habitat for fifteen kinds of zooplankton species in the sampling period; they were *Cyclopid copepoda, Daphnia, Bosmina, Copepoda nauplius, Diaptmus, Kellicottia, Keratella cochlearis, Brachionus, Keratella quadrata, Asplanchna, Filinia, Polyarthra, Notholca, Sida crystallina, and Scapholeberis.* Total number of zooplankton in Bird Sanctuary Pond (6630 Ind·L^{-1}) was the highest. Especially for *Keratella cochlearis*, it represented 5014 Ind·L^{-1} in Bird Sanctuary Pond. The second pond which had higher total number of zooplankton species was Men's Bathing Pond (3534 Ind·L^{-1}); *Kellicottia* dominated the total number of zooplankton in this pond was 2560 Ind L^{-1}. The next two ponds were Model Boating Pond (184 Ind·L^{-1}) and Highgate NO.1 Pond (1734 Ind·L^{-1}). Model Boating Pond had a high number of *Bosmina* (875 Ind·L^{-1}).

3.2.2. Zooplankton species richness
Three main species groups in each pond were *Rotifer*, *Cladocera,* and *Copepod* (Figure.1). The percentage of *Rotifer* group in Bird Sanctuary Pond was up to 90.00%, the next was 80.00% in Men 's Bathing Pond and *Rotifer* group in the whole Ponds Chain was around 75.00%. *Cladocera* occupied the half of species group in Model Boating Pond, and more than 40.00% in the NO.1 Pond. *Copepod* dwelled much in Stock Pond but less in other four ponds.

IOP Conf. Series: Earth and Environmental Science **68** (2017) 012015 doi:10.1088/1755-1315/68/1/012015

The Shannon Diversity was a typical index to represent the species diversity in freshwater biology. Figure. 2 listed the Shannon Diversity for the target five ponds. No.1 Pond had highest Shannon Diversity value of 7.77. However, the lowest reservoir was Bird Sanctuary with value of 1.15.

3.3. Chemical substances in each pond

Highgate No.1 Pond had 13.27 mg·L^{-1} oxygen and the oxygen content was the highest among five ponds. Men's Bathing Pond (9.11 mg·L^{-1}) ranked second, and then was Model Boating Pond (6.35 mg·L^{-1}). The forth one and fifth one were Bird Sanctuary and Stock Pond.

Table 2. The composition of zooplankton species in ehch pond

Species	Stock	Bird	Boating	Men's	No.1
Cyclopid copepoda	105	216	3	97	81
Daphnia	112	4	22	401	266
Bosmina	15	92	875	11	402
Copepoda Nauplius	16	215	61	116	9
Diaptomus	94			5	
Kellicottia		35	209	2560	358
Keratella cochlearis	28	5014	223	165	189
Brachionus			19		18
Keratella Quadrata	12	314	5	20	21
Asplanchna		393	68		
Filinia			168		82
Polyarthra		244	115		5
Notholca	1		48	77	66
Sida crystallina	15	31	6	64	117
Scapholeberis		71	19	18	22
Total Number(ind·L^{-1})	399	6630	1841	3534	1734

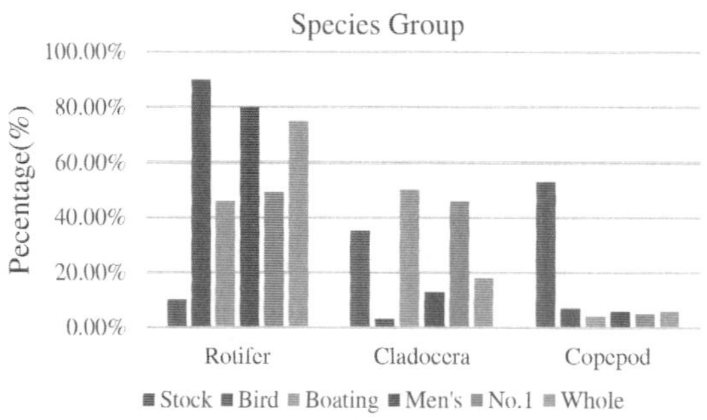

Figure 1. The percentage of three zooplankton groups in each pond

The highest SRP (soluble reactive phosphorus) was in No.1 Pond, the next one was in Bird Sanctuary Pond. The other pond which had SRP more than 100 mg·L^{-1} was Stock Pond. Then, the rest ponds had less than 100 mg·L^{-1} SRP.

Highgate No.1 Pond had the highest total Phosphorus which was 610.09 (μg/L), Bird Sanctuary and Stock Pond had 262.68 (μg·L^{-1}) and 203.72 (μg·L^{-1}) TP respectively. The next two ponds were Boating Pond and Men's Bathing Pond.
Men's Bathing Pond had the lowest content of NN (Nitrate Nitrogen), 0.155 (mg·L^{-1}). Boating Pond, and Stock Pond had more NN. The higher one was in Bird Pond. No.1 pond had the highest content of NN.

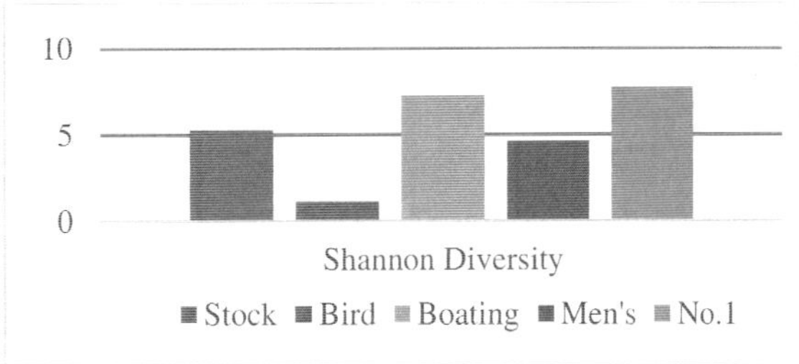

Figure 2. Shannon Diversity for each pond.

Table 3. major pond code in each pond

Pond code	Stock	Bird	Boating	Men 's	No.1
Oxygen (mg·L^{-1})	3.16	5.44	6.35	9.11	13.27
SRP (μg·L^{-1})	152.41	237.38	42.64	14.06	416.33
TP (μg·L^{-1})	203.72	262.68	76.68	52.12	610.09
NN (mg·L^{-1})	0.827	1.008	0.249	0.155	2.256

3.4. Environmental variables
Conductivity caused decrease of *Daphnia* and *Kellicot*. And for *Bosmina*, the number increased initially and then it nearly reached to zero. And *Keratella cochlearis* increased rapidly.

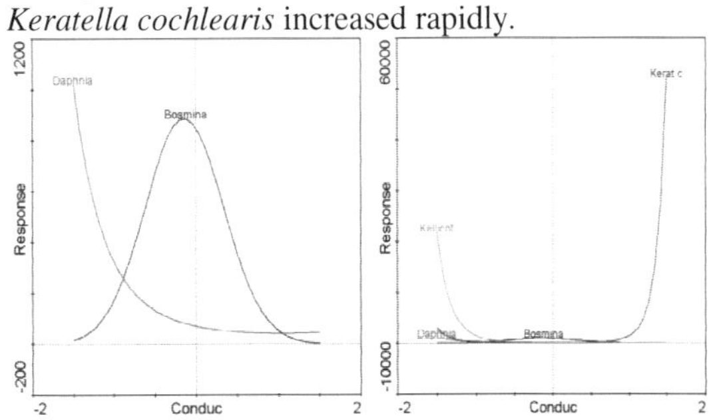

Figure 3. Generalized Linear Models shows the relationship between conductivity and zooplankton.

Total phosphorus had a mainly influence on *Kellicottia*, and made it increase. *Keratella cochlearis* initially increased and then decreased as the total phosphorus increased.

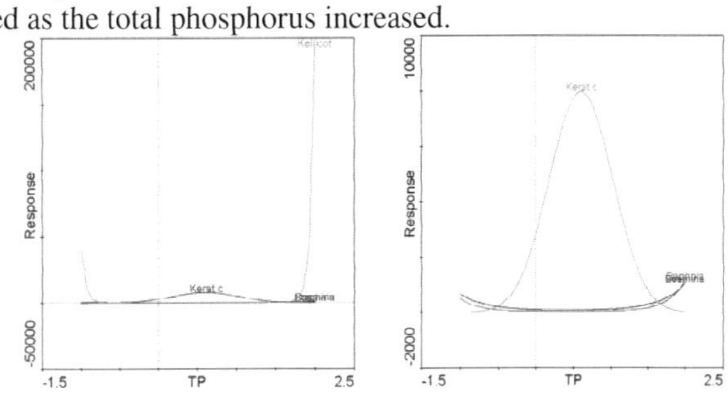

Figure 4. Generalized Linear Models shows the relationship between total phosphorus and zooplankton.

When the number of predation reached around one, *Keratella cochlearis*, *Daphnia* and *Bosmina* fell to zero but *Bosmina* started to rise then. And the curve of *Daphnia* was symmetrical from negative x-axis to positive x-axis. *Kellicottia* increased and then decreased.

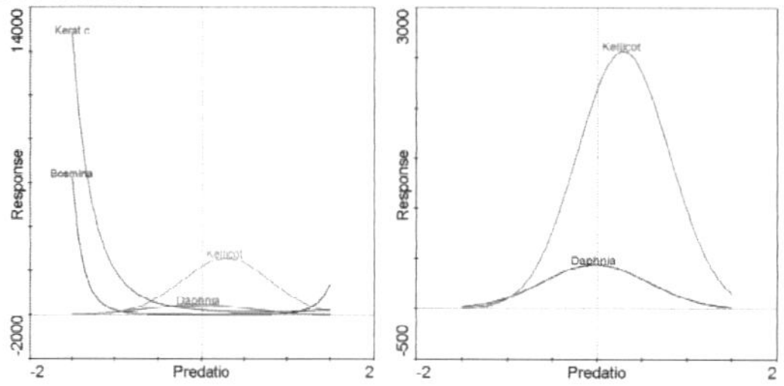

Figure 5. Generalized Linear Models shows the relationship between fish predators and zooplankton

When Chlo α increased, *Bosmina* fell first and then increased. But for *Daphnia*, it rose at first and fell finally.

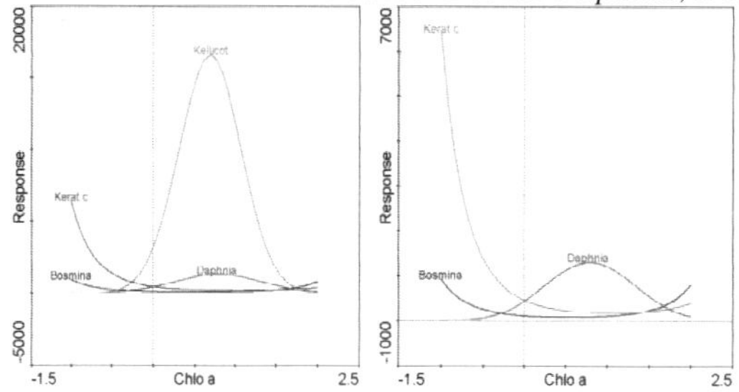

Figure 6. Generalized Linear Models shows the relationship between chlorophyll α and zooplankton.

Keratella cochlearis and *Kellicot* rose firstly and then fell in different time but they all became zero when the number of dissolved oxygen was two. The slope of *Bosmina* and *Daphnia* was upward originally and then downward.

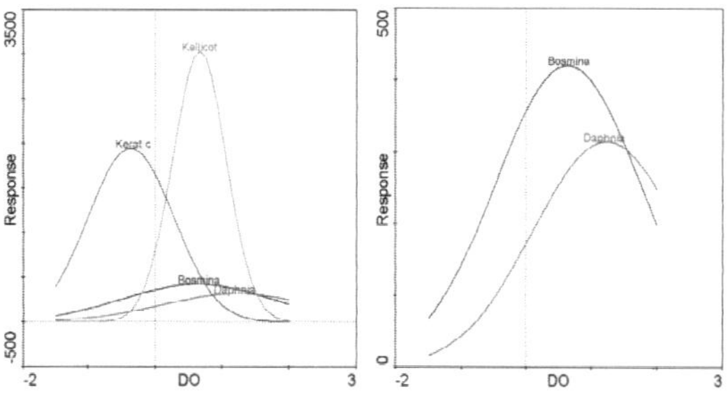

Figure 7. Generalized Linear Models shows the relationship between dissolved oxygen and zooplankton.

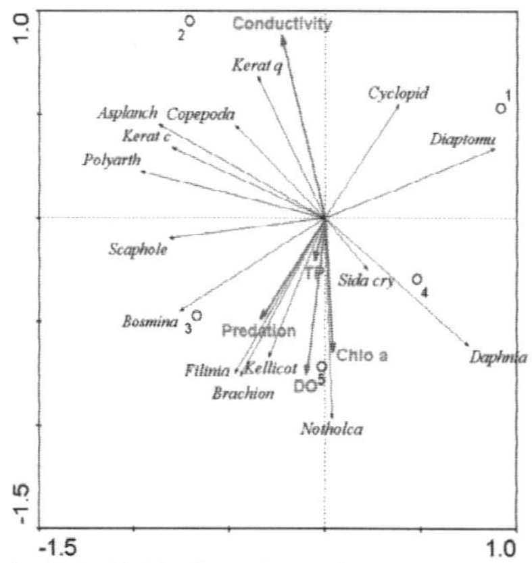

Figure 8. Redundancy analysis (RDA) bi-plot of zooplankton species and environmental variables.

From Fig. 8, conductivity had the positive correlation with *Kerat q* mainly, and had negative correlation with *Notholca*. It was the main factor which could cause a change in zooplankton communities. Predation was proportional to the communities of *Filinia*, *Kellicot* and *Brachion* and inversely proportional to *Cyclopid*. Total phosphorus had relatively less impact on zooplankton communities but also could cause a decrease of *Kellicot* as it fell. The increase of dissolved oxygen might result an increase on *Notholca* or a decrease on *Keratella quadrata* and *Cyclopid*. While chlorophyll α increased, *Notholca* increased extremely. In addition, it almost had no influence on *Scaphole*.

4. Discussion

4.1. Zooplankton species richness and diversity

Generally, two process categories, local processes (e.g., environmental heterogeneity and species interactions) and regional processes (e.g., dispersal, connectivity), are the main factors which can influence the community composition and species diversity [12]-[14]. Beside the general impacts, local factors have effects on zooplankton species richness which have been determined: lake area and primary productivity [15], [16], [10], water quality [17], [18], lake depth [19], latitude [20], nutrients [17], [21], predation and competition [22].

From the result, the data showed that zooplankton communities in each pond were highly fluctuated. Reference [23] indicated that within the finite range in habitat sizes of the ponds studied, the number of species had no statistic relationship with the ponds area. In Highgate Pond Chain, the relationship between the ponds area and zooplankton communities was not distinct. The largest area in the target chain was Men's Bath-18250 m^2 and the smallest one was 4401 m^2. The number of zooplankton in Men' s Bathing Pond was relatively high, and that in Stock Pond, the smallest zone, was dramatically tiny refer to the study. Based on the field observation, Bird Pond, the pond with a plenty of vegetation coverage on and around it, had richest species. However, Bird Pond held an extreme difference in aspect of area to Men's Bath Pond. Therefore, zooplankton communities were not correlated with area obviously.

Refer to the local geography, Stock Pond was in the highest elevation of 87.97m, and No.1 Highgate Pond sited in 65.6m. However, the total number of zooplankton communities in Stock Pond was the least, and remarkably less than No.1 Highgate Pond. Thus, the elevation in Highgate Pond Chain is not the factor which can negatively influence the zooplankton in local. In addition, the elevation difference between each pond in this study is not remarkable, so that the elevation of each pond technically has no marked influence on the zooplankton communities. The tiny difference in elevation typically could be ignored during the biological study. Water body depth was regarded as a fundamental index in biological studies as well. According to the basic information of Highgate Pond Chain, Highgate Men's Bathing Pond was the deepest one (>2.5m), and Stock Pond was the shallowest one among the five ponds. However, the depth of these ponds was not related to the zooplankton species richness, the species richness in these two ponds were both relatively high from the result, and statistical correlation was not obvious.

The Highgate pond field work records showed that the Stock Pond had many branches and leaves covered on water, and the de-composition activity was relatively vibrant. This was the primary factor causing the great consumption of dissolved oxygen. Therefore, this reason shall be placed as major account of zooplankton community deficiency. Only a few bankside vegetations could be found around Model Boating Pond and the banks were entirely re-enforced. Therefore, the solar accessibility in Model Boating Pond was considerable. In contrast, Men's Bathing and Stock Ponds had enormous crown coverage above the water body. Hence, the sunlight was not available to some extent in these two ponds. Correspondingly, *Bosmina* was richest in Boating Pond and scarce in Men's Bathing and Stock Ponds. As a result, the surrounding of each pond and the solar accessibility could positively impact on the species richness, in this study area.

Diversity has been regarded as a significant indicator in biological survey, usually focused on species richness and distribution. Three possible explanations were considered to illustrate the low diversity: low richness and uniform distribution, high richness but with dominate species and low richness with dominated species. Mathematically, nearly all 13 zooplankton species were uniformly distributed in No.1 Highgate Pond. Therefore, the Shannon Diversity for this pond was dramatically larger than others. The Shannon Diversity was the lowest in Bird Sanctuary Pond; however the total number of zooplankton species was the highest. One remarkable factor that could determine the low Diversity parameter was the extreme dominance species- *Keratella cochlearis*.

4.2. The relationship between environmental variables and the zooplankton communities in Highgate Pond.

RDA analyses showed that conductivity, dissolved oxygen, chlorophyll α were the most significant environmental variables in Highgate ponds. Reference [24] performed a research and found that 41 percent of the variance in species richness could be contributed to conductivity. Conductivity was positively correlative with eight zooplankton species and negatively related to seven species, and especially it had the largest influence on species. Literally, conductivity was regarding to the number of removable and accessible ions in the ponds. Some zooplankton species were able to adapt the aquatic environment easily but some were not. On the basis of the result (Figure.3), when the value of conductivity rose, the number of *Kellicot* and *Keratella cochlearis* was altered. Since *Kellicot* decreased extremely, this species was vulnerable against the conductivity, and also *Kerat c* increased exceedingly, the tolerance was smaller than *Kellicot*. In addition, *Daphnia* fell as *Bosmina* increased in a certain interval that the conductivity maintained in low numeric values. These two zooplankton species had a typically competitive relationship to each other.

Dissolved oxygen ranked second in influencing the communities of zooplanktons, and this index was able to drive eight species in positive direction, as well as seven species in negative correlation. This parameter stood opposite to the conductivity in RDA analysis. Species *Keratella cochlearis* increased initially and decreased with the richness of dissolved oxygen. Dramatically, species *Kellicot* increased rapidly when the community of *Keratella cochlearis* was decreased. Therefore, the most reasonable explanation was the competitive relationship between these two species within this certain ecological environment. Even *Bosmina* and *Daphnia* had a less amount; the tolerance of them was nearly same as that of *Keratella cochlearis*. The peak of *Kellicot* was overly high in species; however its environment tolerance was comparatively lower than others. The dissolved content for oxygen in Bird Sanctuary Pond was the most suitable for *Keratella cochlearis* to survive, since DO in this pond was not extremely high or low.

Predation also was a factor which can impact zooplankton species, whereas the influence was not significant, because of the intensive competition between species. For example, the competition between *Daphnia* and *Bosmina* also existed. Past studies suggested that fish predation could impact the density of zooplankton; however, most studies had not clearly presented the influence in details. Same scenario could be found in Highgate Pond chain.

In Highgate pond chain, the chlorophyll α distribution was disordered without any linear relationships. Chemically, Dissolved Oxygen content was dramatically important to the trend of Chlo a. Highgate No.1 pond was characterized by high density of algae and plankton (high chlo α).*Kerat c* was extremely increased as the content of Chlo α had been raised; therefore the tolerance community change of the species against Chlo α was not obvious. From figure 3, the comparative relationship between *Daphnia* and *Bosmina* could be extracted as well, and the tolerance of them was identity and relatively in a high level. In Boating Pond, the content of Chlo α was ranked in the middle. However, *Bosmina* in the pond was the highest among these five ponds, and *Bosmina* was the dominated species in this target pond. This result was different to the relationship with chlo α, and the reason could be explained by other factors, such as the competition, nutrients and physical phases. Therefore, low dissolved oxygen and primary productivity restricted the zooplankton composition, richness and diversity. Whereas, the data maintained that these two factors could not determine the composition of zooplankton. As a result, the matrix of environment in Highgate pond chain was complex and disordered.

Nutrient, for an outstanding example: total phosphorus, was unable to affect species intensively. The content level distribution within these five ponds was obviously. Boating and Men's had dramatically low TP content, in contrast, the TP content in other three reservoirs were visually high. For stock pond, the deposition of TP was regarded as main activities of the biological nutrient activities. For bird and no.1, the biological activities were obviously extreme, and the nutrients physical input was relatively higher than others. According to the local survey, Boating and Men's Bathing Ponds were maintained by anthropogenic activities. Park volunteers cleaned up these two ponds, by algae salvage. TP was inversely related with nine species and positively related with six species as per the biological relationships. Refer to Figure 4, *Kellicot* was the specie that prefer to high content of TP, it could reach its peak response in a tiny content range so that the tolerance of it was exceedingly sensitive. *Daphnia* and *Bosmina* were in the same drift as the TP changed, the response of the community to these two species was quite gently, therefore, the environmental feedback was not sensitive. However Men's Bathing Pond was lack of *Bosmina* but was relatively sufficient with *Daphnia*, thus the comparative relationship was obvious in the Men's Bathing Pond as well.

Environmental variables in Highgate pond chain showed the extreme impacts on zooplankton communities from the statistical analysis and previous academic survey. Conductivity, dissolved oxygen, total phosphorus, chlo α and predation maintained that the influence were in different extent. Some variables were significantly correlated to gradient of the chain. Some factors showed that the gradient correlation was not apparent, as they were derived by artificial activities.

5. Conclusion

This study clearly indicated that conductivity, chlorophyll α, dissolved oxygen and predation can be important factors to influence zooplankton communities, including species richness, diversity and distribution, in Highgate Chain, London. Conductivity could be regarded as the most significant indicator to alter the ecological structure of zooplankton. As each pond has different physical and chemical environmental variables, the zooplankton communities were generally in high diversity. Some variables showed highly influence on zooplankton distribution in other papers were not important here.

6. Reference

[1] R. M. Baxter (1977). Environmental effects of dam and impoundment. *Annual Review of ecology and systematics*. 8, 255-283.
[2] K. A., Fahd, M. Arechederra, , D. Florencio, Leon, & Serrano, L. (2009). *Copepods and branchiopods of temporary ponds in the Doñana Natural Area (SW Spain)*: a four-decade record (1964–2007).
[3] S.R. Carpenter, N. F. Caraco, D. L. Correll, R. Howarth, W., Sharpley, A. N., Smith, V. H. (1994). NONPOINT POLLUTION OF SURFACE WATERS WITH PHOSPHORUS AND NITROGEN. *Ecological Applications*.
[4] D. M. Frederick, (1998) *The Uncommon Guide to Common Life on Narragansett Bay*. Save The Bay.
[5] R.G. Wetzel, (2001). Limnology: Lake and River Ecosystems. Th ird edition. (pp.1006) Academic Press, San Diego, London.
[6] D. G. George, (1991). *The influence of Global Warming on Freshwater Plankton Communities in Britan.* Institute of Freshwater Ecology, Windermere Laboratory, Far Sawrey, Ambleside, Cumbria LA22 OLP, England.
[7] J. Sellami, A. Elloumi, A. Hamza, M. Mhamdi, & H. Ayadi, (2011). *Local and regional factors influencing zooplankton communities in the connected Kasseb Reservoir,* Tunisia. Journal Home, Volume 37, No 2.
[8] A. P. Allen, T. R. Whittier, P. R. Kaufmann, *et al.* (1999). Concordance of taxonomic richness patterns across multiple assemblages in lakes of the northeastern United States. *Can. J. Fish. Aquat. Sci.,* 56, 739–747.
[9] A. P. Allen, T. R. Whittier, P. R. Kaufmann *et al.* (1999). Concordance of taxonomic composition patterns across multiple lake assemblages: effects of scale, body size, and land use. *Can. J. Fish. Aquat. Sci.,* 56, 2029–2040.
[10] S. Dodson, S. Aenott, & K. Cottingham (2000), The relationship in lake communities between primary productivity and species richness. *Ecol.* 81, 2662-2679.
[11] T. W. Lewis, J. C. Makarewicz, & P. Bertram, (1998). Changes in Phytoplankton Size-class Abundance and Species Composition Coinciding with Changes in Water Chemistry and Zooplankton Community Structure of Lake Michigan. *Great Lakes Res.* 24(3):637-657. Internat. Assoc. Great Lakes Res.,
[12] R. Ricklefs, 1987, Community diversity: relative roles of local and regional processes. *Science* 235, 167-171.
[13] N. Mouquet, & M. Loreau, 2003, "Community patterns in sourcesink metacommunities," *Am. Nat.* 162, 544-557.
[14] M. Holyoak, M. Lwibold, & R. Holt, 2005, *Meta communities: Spatial Dynamics and Ecological Communities.* Chicago University Press, Chicago.
[15] S. Dodson, 1991, Species richness of crustacean zooplankton in European lakes of different sizes. Verhandlungen Internationale Vereinigung für Theoretische und Angewente Limnologie 24, 1223-1229.
[16] S. Dodson, 1992, Predicting crustacean zooplankton species richness. Limnol. Oceanogr. 37, 848- 856.
[17] E. Jeppesen, J. Jensen, M. Søndegaard, T. Lauridsen, & F. Landkildehus, 2000, Trophic structure, species richness and biodiversity in Danish lakes: changes along a phosphorus gradient. *Freshw Biol* 45:201–218.
[18] K. Cottenie, E. Michels, N. Nuytten, and L. De Meester, 2001, Zooplankton community structure and environmental conditions in a set of interconnected ponds. *Hydrobiologia* 442, 39-350.
[19] W. Keller, and M. Conlon, 1994, Crustacean zooplankton communities and Lake Morphometry in Precambrian shield lakes. *Can. J. Fish. Aquat. Sci.* 51, 2424-2434.
[20] P. Hebert, and B. Hann, 1986, Patterns in the composition of arctic tundra pond microcrustacean communities. *Can. J. Fish. Aquat. Sci.* 43, 1416-425.
[21] M. Leibold, 1999, Biodiversity and nutrient enrichment in pond plankton communities. *Evol. Ecol. Res.* 1, 73-95.
[22] J. Brooks and S. Dodson, 1965, Predation, body size, and composition of zooplankton. *Sci.* 150, 28-35.
[23] C. Karl, E. Michels, L. Neys, and L. De Meester, 2001, Zooplankton on the move: first results on the quantification of dispersal of zooplankton in a set of interconnected ponds. *Hydrobiogia*, January 2001, Volume 442, Issue 1, pp 117-126
[24] U.T. Hammer, andL. Forró, 1992, Zooplankton distribution and abundance in saline lakes of British Columbia, *Canada.International Journal of Salt Lake Research* 1:65-80.

Research on Estimates of Xi'an City Life Garbage Pay-As-You-Throw Based on Two-part Tariff method

Shi Yaobo, Zhao Xinxin and Zheng Fuli

Xi'an University of Technology, School of Economics and Management, Xi'an, Shaanxi, 710054

Abstract: Domestic waste whose pricing can't be separated from the pricing of public economics category is quasi public goods. Based on Two-part Tariff method on urban public utilities, this paper designs the pricing model in order to match the charging method and estimates the standard of pay-as-you-throw using data of the past five years in Xi'an. Finally, this paper summarizes the main results and proposes corresponding policy recommendations.

1. Introduction

With the rapid improvement of the level of urbanization in China, the problem of municipal solid waste becomes more and more serious, and the funds needed for the effective treatment of municipal solid waste are increasing, resulting in heavy financial burden. How to control the emissions from the source to achieve the dual objectives of environmental governance and alleviating burdens on finance, which is a major issue in front of public policy makers. Referring to the international experience, the garbage charge system, which is widely used in the world, is an important economic means to control and reduce the pollution from the source. At the beginning of 2002, the Ministry of Construction issued " notice on the implementation of city life garbage fee system to promote garbage disposal industrialization "and "city garbage management measures", requiring that according to the actual situation of the region and local conditions, waste collection system should be implemented scientifically and municipal solid waste charging standards should be formulated. In September 2002, the state promulgated the "opinions on promoting urban sewage and the development of garbage treatment industrialization", and further put forward that the garbage disposal fee should be levied according to the local actual conditions. But now many city garbage fees levied are in different degree, most cities charged fees at a fixed price, and even at present, there is still some cities don't start urban living garbage disposal fee [1]. Therefore, calculating the reasonable price of garbage is the premise to carry out pay-as-you-throw schemes.

At present, most countries in the world mainly adopts fixed charges and pay-as-you-throw city life garbage charge method [2].The fixed charges is suitable for the initial stage, but when the garbage fee develops to a certain level, it's effect is very limited. Therefore, it is necessary to implement the pay-as-you-throw system, which can achieve higher returns with low cost. At present a fixed charge model of Xi'an city although to some extent relieved the situation that the funds to run the garbage disposal facilities are inadequate. But in the long run, it can't change the concept of garbage emissions and achieve environmental goals. Many foreign countries and regions achieved good results by pay-as-you-throw policy, reducing the waste output and making garbage collectors reduce the waste emissions [3]. Therefore, in the long run, pay-as-you-throw policy should be measured in cities. The research on the pricing of domestic waste charge has been relatively rich, but the domestic research on the pricing of garbage collection is almost blank. Therefore, this paper uses the two pricing method to calculate the price of urban garbage, to determine the reasonable price and to provide a reference for the promotion of urban garbage management capacity.

2. The choose of two-part pricing method

Choosing a pricing model to match the pay-as-you-throw is the key to set charging standards. The following two points about pricing models need to be payed attention: first of all, the pricing model need to connect the municipal solid waste charging with the actual life garbage emissions; secondly, the pricing model needs to make up for the treatment cost of municipal solid waste disposal site. The cost of living garbage disposal facilities are divided into fixed investment cost and variable operating costs by two part pricing method. It's basic fee is based on the fixed investment in the early stage of garbage disposal facilities, or we can say that it is based on initial input costs and capital gains, which has nothing to do with the actual discharge of domestic waste. Fees according to quality are variable operating expenses, which are based on the actual amount of domestic waste emissions. The application of the two-part pricing method in natural monopoly industry can be traced back to the earliest 1892 when Hopkinson in British designed the electricity price system. After that, it was be perfected formally by Lewis and Willing [4].The advantage of the two-part pricing method is that it is based on the balance of payments to realize economic welfare maximization, which has strong maneuverability. At the same time, it is based on marginal cost pricing, and actually it is a kind of optimal pricing. But it is also limited to the assumption that the market information is complete and there is no other transaction cost. Although there are still some shortcomings in the implementation of the two part pricing method, the introduction of the two pricing method has its unique advantages:

Firstly, that separating the basic fee from the unit price can more effectively compensate for the cost and ease the problem of domestic waste emissions.

Secondly, two-part pricing method can be developed for different regions of different charges, achieve the specific analysis of specific issues, and help to achieve a reasonable allocation of resources.

3. The price calculating of living garbage for pay-as-you-throw

3.1. Model construction

Rubbish disposing costs include the cost of distribution, such as waste transportation, transfer, handling costs, and the cost of public sector management, such as the cost of collection, monitoring costs and education costs. Municipal domestic refuse treatment fee can be divided into fixed cost and variable cost in this two pricing model.

After the above analysis, combined with the principle of Two-part tariff models, we can create model for the unit garbage fee, such as the model shown in equation1:

$$T = X + Y \tag{1}$$

Wherein X is the fixed cost in solid waste disposal fees, the garbage fee is zero as well as no amount of garbage generated, then the fixed cost is calculated as follows:

$$Q < Q_1, T = X; \quad X = \frac{C_1}{Q_1} \tag{2}$$

The C_1 is fixed cost of landfill and the Q_1 is average daily production of municipal solid waste planned in Equation2.And Y is the flow of solid waste disposal fees also the cost of incremental costs. It is calculated as:

$$Y = \frac{(1+R)C_2}{Q-Q_1} \tag{3}$$

The C_2 is the total operating costs in Equation 3, which including staff salaries and benefits , other labor costs, various office expenses, operating expenses, depreciation of fixed assets, equipment maintenance fees, etc in garbage disposal plant. The household garbage of more than optimal life rubbish emissions will collect unit cost:

$$T = \frac{C_1}{Q_1} + \frac{(1+R)C_2}{Q - Q_1}$$

(4)

After the pricing model of unit garbage has designed, we can based on the actual emissions from specific residents to charge amount-based by the residents within a certain time. So we need to understand the amount of waste per emissions. Let Z be the garbage fees for people's actual payable, to give the formula:

$$Z = \begin{cases} \frac{C_1}{Q_1}, & Q_2 < Q_3; \\ \frac{C_1}{Q_1} + \frac{(1+R)C_2}{Q - Q_1}, & Q_2 > Q_3 \end{cases} \quad (5)$$

The model 5 reflects if the residents want to reduce the solid waste disposal fees payable, you will need to reduce emissions of garbage. So not only can pay less solid waste disposal fees, but also to reach the target of the effect of garbage reductions.

3.2. Parameter Settings and Setting-up of Fees Standard

According to bulletin of statistics for the bureau of Xi'an landscaping in 2015, listing 1:

Table 1. 2010--2015 Statistical Data

years	2010	2011	2012	2013	2014	2015
Production of Refuse(10000 tons /year)	233	237	265	287.32	290.76	359.37
Production of Refuse(10000 tons /day)	6316.14	6493.15	7260.27	7871.78	7966.03	9845.75
Cleaning area (10000 square meters)	5900	6290	6411	6952	8725	10110
Sanitation vehicles (cars)	1042	1042	1200	1279	1639	1990
Domestic garbage clean-up rate (%)	93.8	93.9	93.7	94.93	93.95	93.48

Data sources: Statistical bulletin of Xi'an Landscaping Bureau in 2015

First, according to the Xi'an Bureau of Landscape Survey data show that currently more than 90% of the Xi'an city garbage removal and transportation to be harmless landfill and buried in River Village Ditch garbage treatment plant, so this article assumes that all of life garbage with innocuous processing in there, this landfill has planned to invest 150 million in the entire project [5]. The plant can be designed to use 2,000 tons at the average daily treatment capacity for 50 years by River Village Ditch landfill, which has been used for 20 years. As the table shows, the Production of Refuse is rapid growth per day with the increase of urban population and urbanization rapid growth in 2010 to 2014, and from the cleaning area of 62.9 million square meters in 2010 to grow to 10,110 square meters in 2014. If it is assumed using a variable rate of charging system when the waste quantities are shown a downward trend, and this may wish to use the average of the past five years to make a clear prediction of the daily average volume of garbage in the future , so $Q^* = \sum_{i=1}^{n=5} X_i / 5 = 7887.40$ ton. While the Average daily processing capacity of only 2,000 tons at landfill in the early stages construction, meaning that it has reduced the time for the estimated time of 1/4, as 7.5 years. There fore,

The fixed costs with used for 20 years: = 15,000 / 50 * 20 = 600 million yuan,

The fixed costs of remaining 7.5 years: = 15000-60000 = 900 million yuan.

Second, the parameter setting. R is the appropriate annual profit margins of garbage disposal Enterprises, herein R is determined with reference to the relevant waste listed companies net profit margin of main business, which was about 7%; The net profit of Xi'an High-tech park and the economic development zone relates to garbage treated about 14 companies net profit margin (including landfill gas power plant in Xi'an, a comprehensive treatment plants and hazardous waste disposal plant), with an average of 6%; Therefore, this article takes on margin R = 6% [6].
Third, We have to organize the data provides by Environmental Sanitation in our city, by the end of 2015, the survey area have households 2.506 million who are permanent and non-permanent residents. Among them, the collector of waste by the streets office is 1.6 million and by the property company is 906000.
The following is three link cost and relevant taxes of accounting domestic waste collection, transportation, processing.

Table 2. 2015 life garbage disposal cost accounting in Xi'an tables

Units: ten thousand yuan

Link / department	Sanitation office	street office	Jiangcun ditch landfill	total	unit cost
Collecting costs	6,500.00	10,240.00		16,740.00	46.58
transportation cost	16,500.00			16,500.00	45.91
Landfill, incineration costs			9,020.00	9,020.00	25.10
Subtotal	23,000.00	10,240.00	9,020.00	42,260.00	117.59
Taxes and surcharges				2,600.00	7.23
total	23,000.00	10,240.00	9,020.00	44,860.00	
Unit cost (yuan)	64.00	28.49	25.10		124.82
Annual waste (ten thousand tons)				359.37	

As can be seen from the above table, the total cost of garbage disposal in Xi'an urban area is 448.6 million yuan, the cost per unit ton is 124.82 yuan. Which collects the cost of 46.58 yuan per ton, the cost of transport 45.91 yuan per ton, incineration costs 25.10 yuan per ton, taxes and additional charges on behalf of the charge fee of 7.23 yuan per ton.

3.3. Based on the two-part pricing method of measurement fees

The total cost of garbage disposal with the amount of garbage emissions, cleaning area increased in Xi'an, to the 2015 data as the base, assuming that the operating costs and garbage volume was a linear relationship, then when the garbage collection capacity of 3.5937 million, the total operating expenses of 448.6 million yuan, the average operating cost per ton of 124.82 yuan.
Came to the conclusion that the total operating expenses in the last seven years. If the reduction in the amount of waste can also reduce the total operating costs, then we take the average of nearly seven years of total operating costs as a calculated reference value.

which is: $\overline{C}^* = \dfrac{x_1 + x_2 + \ldots + x_n}{n} = 33685$ ten thousand yuan

If you use the general linear type, calculate the total operating costs of the year, the list is as follows:

Table 3. Total Operating Cost Calculation

years	2009	2010	2011	2012	2013	2014	2015
x(10000 tons or 10000 square meter)	179	233	237	265	287.3	290.7	359.37
$c = bx$ (ten thousand yuan)	22342.78	29,083.06	29,582.34	33,077.30	35,860.79	36,285.17	44,856.56

Therefore, the total operating expenses of the treated garbage are set at 336.85 million yuan. It can get a fixed cost of living garbage treatment fee in Xi'an:
Incremental charges are, take 2015 as an example:

$$X = \frac{C_1}{Q_1} = \frac{90000/7.5/12/30}{0.33} = 101\text{yuan/ton}$$

$$Y = \frac{(1+R)C_2}{Q\text{-}Q_1} = \frac{1.06 \times 33685}{359.37 - 118.8} = 148.42\text{yuan/ton}$$

Pricing standard for unit life garbage collection is: T=X+Y=249.42 yuan/ton$\approx$0.25 yuan/kg.
Known in 2015 the number of inhabitants was 2 million 502 thousand and 600, the optimal waste emissions per household was 472 kg (118/250.26*1000), that is, when the household emissions are within the 472 kg, they only need to pay the fixed cost of 47 yuan (101*472/1000), equivalent to 0.1 yuan per kilogram. When the household emissions are more than 472kg, the users need to pay the cost of 0.25 yuan / kg. In summary, through the pay-as-you-throw model, government can give the residents certain subsidies in the different stages to achieve the purpose of appropriate relief and benefits according to the specific development of charging system.

4. Conclusion and recommendations

In this paper, the two-part pricing method is used to calculate the price of Xi'an city household garbage, we can draw the following conclusions from this paper:

Pay-as-you-throw schemes of household garbage mainly consider the fixed costs and variable costs, further reflecting the fairness of cost allocation and maximizing the interests of users. At the same time, according to the pay-as-you-throw principle, users who produce more garbage need to pay more, and this interest orientation helps to form city residents' consumption concept and ultimately develop the green consumer preferences, encourage enterprises to cleaner production and ultimately reduce waste emissions to achieve a resource-saving and environment-friendly society.

In order to use the two-part pricing method to calculate the price of Xi'an municipal solid waste, this paper puts forward the following policy recommendations:

First, determine a reasonable fee standards. The establishment of garbage charge standard faces many difficulties: on the one hand, it is difficult to accurately calculate the cost of waste disposal; on the other hand, other factors determining the cost of fees (such as sanitation management costs, the price of labor) lack of unified standard basis. Therefore, Xi'an should be based on it's actual economic and social development, calculates and determines a reasonable fee, in order to ensure the city life garbage fees can be really carried out, and the two-part pricing method can play a better role.

Second, take effective measures to ensure the full collection of garbage disposal fees. Appropriate garbage collection mechanism should be established to improve the efficiency of garbage collection and charging rate. The current charging method of garbage collection are: automatic payment, commissioned collection and site payment, and most people choose site payment, which increases the cost of charges. Therefore, the Xi'an municipal government should consider open more channels to collect the fees or entrust some corresponding charging institution to collect garbage disposal fees.

Coexistence of a variety of charge modes can improve the efficiency of the charges as well as the convenience of urban residents.

5. References

[1] Chen Minxia. Research on the pricing of municipal solid waste charging [J]. Jiangsu Environmental Science and Technology, 2008, (S1):24-28.
[2] Zhang Xuecai, Li Dayong. Comparison of municipal solid waste charging [J]. Ecological economy, 2005, (10):98-100.
[3] Chen Minxia. Comparison of domestic and international urban domestic garbage charging experience [J]. Economist, 2008, (06):92-93.
[4] Wang Qiuyue. Research on the pricing mechanism of natural monopoly industry [M].Nanjing: Nanjing University Press, 2011:36.
[5] Wang Yingying. Study on the influence of Xi'an Jiang cungou landfill leachate on the surrounding ecological environment in Xi'an City [D].Xi'an, Chang'an University,2008.
[6] Cao Na． Study on the price of domestic refuse disposal in China [D].Beijing: China University of Geosciences, 2010.

Resource Assessment of Tidal Current Energy in Hangzhou Bay Based on Long Term Measurement

Feng Zhang[1], Chun-Ni Dai[2], Xue-Feng Xu[1], Chuan-Kun Wang[1], Qin Ye[1]

[1] Second institute of oceanography SOA, 36 Baochubei Road, Hangzhou, China
[2] Zhejiang design institute of water conservancy & hydro-electric power, 66 Funingxiang, Hangzhou, China

E-mail: xuxuefeng1981@163.com

Abstract: Compared with other marine renewable energy, tidal current energy benefits a lot in high energy density and good predictability. Based on the measured tidal current data in Hangzhou Bay from Nov 2012 to Oct 2012, this paper analysed temporal and spatial changes of tidal current energy in the site. It is the first time measured data of such long time been taken in tidal current energy analysis. Occurrence frequency and duration of the current of different speed are given out in the paper. According to the analysis results, monthly average power density changed a lot in different month, and installation orientation of tidal current turbine significantly affected energy acquisition. Finally, the annual average power density of tidal current energy with coefficient Cp in the site was calculated, and final output of a tidal current plant was also estimated.

1. Introduction

Tidal current energy refers to kinetic energy in horizontal movement of seawater under the action of tidal force. The kinetic energy can be transformed into electric energy through tidal current turbine. Tidal current energy has a unique superiority in that it does not produce any waste as clean renewable energy. Compared with other marine renewable energy, tidal current energy benefits a lot in high energy density and good predictability, etc. China has conducted survey and assessment on national tidal current energy [1], [2], as well as assessment of tidal current energy in certain regions [3], [4]. These assessments are mainly based on short-term observations or 2-D numerical results, whose purpose is to understand resource conditions in a larger range, and provide a reference for site selection of tidal current energy. It is the first time measured data of such long time been taken in tidal current energy analysis. It will be useful in tidal current turbine design and economic analysis for tidal current energy extraction.

At present, the study of tidal current turbine is becoming the focus of tidal energy research [5], [6]. which need detailed analysis of tidal current characteristics and temporal and spatial distribution at alternative tidal energy extraction site. On the one hand, it is important to choose laying depth and installation orientation of the turbine; on the other hand, it is to grasp occurrence frequency and duration of the current of different speed through long term measurement, so as to better determine design current velocity for tidal current turbine, and estimate its output precisely.

Hangzhou Bay is known for its strong tide, and its energy density of tidal current is one of the highest in China. In this paper, we analyzed from the above two aspects with one year data measured in the central area of Hangzhou Bay, to provide reference for tidal current turbine design, estimation of generating hours of high flow velocity and output of single set.

2. Data source

The data used in this paper was provided by a bottom-seated observation system located in central area of Hangzhou Bay where water depth is about 10m. Specific location is shown in Table 1 & Figure 1. The system was installed with a 600kHz Acoustic Doppler Flow Profiler (ADCP), which emit sound waves upwards to measure flow velocity, flow direction from 1 to 9m above seabed with layer thickness of 1 m. And it recorded velocity and direction date every half hour from November 2012 to October 2013.The accuracy of the ADCP measured velocity is 1% of measured value, and the accuracy of direction is ±1°. The velocity range is ±10ms-1, and the direction range is 0 to 360°.

Table 1. Location of the observation system in Hangzhou Bay

	Longitude	Latitude	Water depth
Station A	121°28.2′E	30°28.6′N	10m

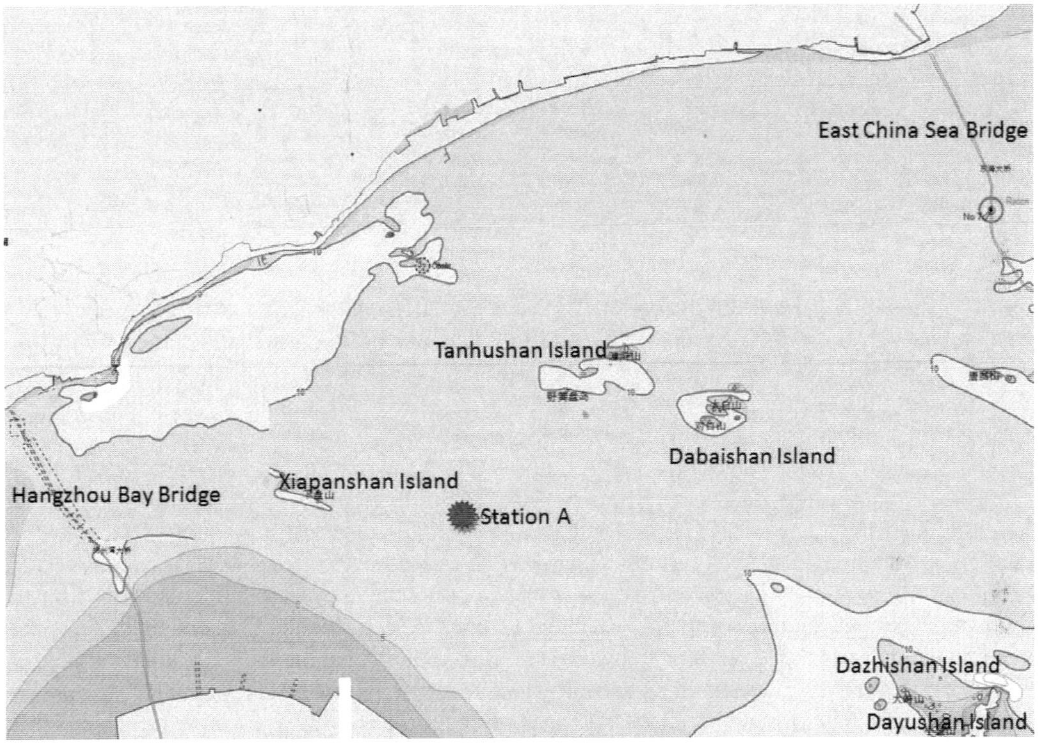

Figure 1. Location of the observation station

At present, most current turbine was designed to be installation a fixed depth below the lowest tide level [5], [7]. That means the surface velocity is usually not used during high tide period. Therefore, the analysis in this paper focused on the current energy of the layers below the lowest tide level, which is in the range of 1m to 6m above the seabed in this site. We call those effective layers.

3. Analysis and discussion

3.1. Basic characteristics of tidal current in the area

The basic characteristics of the tidal current in the station are as follows: The tidal current of Hangzhou Bay is semidiurnal tide. And it has a feature of standing wave for its relation between tide and tidal current. The maximum current velocity basically appears near half-tide level time; while turn of tidal current mostly occurs at high and low tide moments (Figure 2). The flood current direction is about 300°, while ebb current direction is about 120°. Both flood and ebb currents have concentrated directions, and the feature of reciprocating current is in evidence.

Table 2. Hours of depth average velocity at all levels

	U≤0.5m/s	0.5<U≤1.0m/s	1.0<U≤1.5m/s	1.5<U≤2.0m/s	U>2.5m/s
hours	1631	2096	2949	1731	353
Proportion	18.6%	23.9%	33.7%	19.8%	4.1%

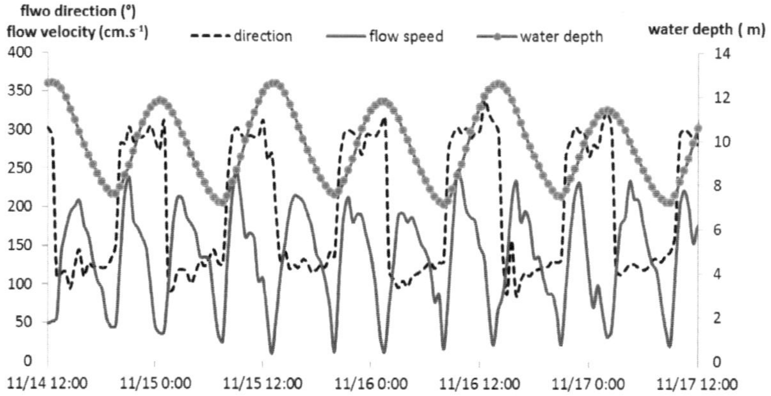

Figure 2. curves of tide and tidal current (September 14 ~ 17, 2012)

During the observation period (September 2012 to October 2013), the maximum flood current velocity in effective layers(1 to 6m above seabed) was 2.73 m/s at 6m layer; the maximum ebb current velocity in effective layers was 2.55 m/s at 6m layer. The maximum depth-average flood velocity is 2.32 m/s, and the maximum depth-average ebb velocity is 2.10 m/s. Flood current lasted an average of 5 hours 55 minutes, while ebb current lasted 6 hours 30 minutes.

During the observation time, depth-average residual value was 0.11m/s, and the direction was about 130 °. The residual value decreases from the surface to the bottom, the direction is deflected counterclockwise, and the residual value near seabed is 0.02m/s, the direction is about 100 °.

Figure 3 shows depth average velocity rose based on measured data of one year. It can be seen that tidal current of the site is mainly concentrated in WNW, while ebb current is mainly concentrated in ENE direction. Flow velocity is mainly in the range of 1.0~1.5m/s, which accounts for 33.7%, followed by the range of 0.5~1.0m/s, which accounts for 23.9% (Table 2).

Flow velocity of one year is analyzed statistically, and velocity value of each guarantee rate is shown in Figure 4. As can be seen, flow velocity beyond 0.5m/s occurred in 7010 hours, with guarantee rate of 80%; flow velocity beyond 1m/s occurred in 4406 hours, with guarantee rate of 50%; flow velocity beyond 1.5 m/s occurred in 1113 hours, with guarantee rate of 13%.

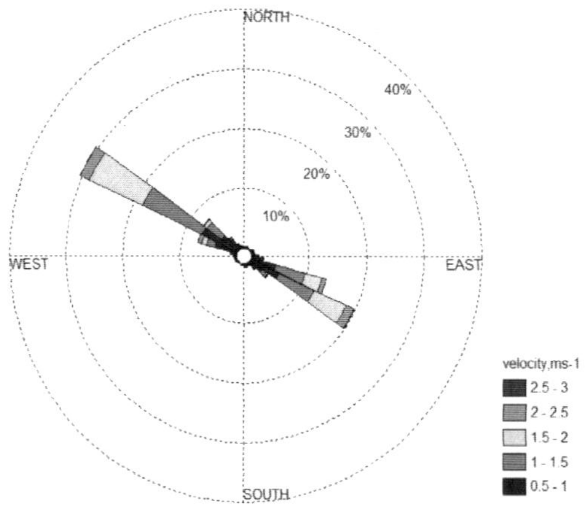

Figure 3. Roses of depth average Tidal current over a year

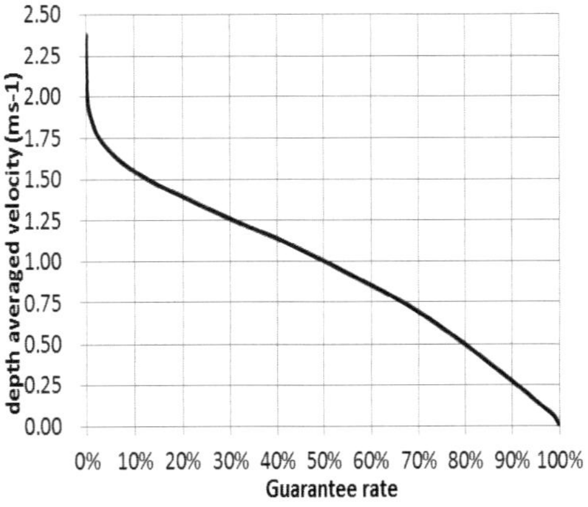

Figure 4. flow velocity for each guaranteerate

3.2. Vertical distribution of tidal current energy

As we all know, tidal current velocity shows significant vertical variation. During the observation, flow velocity of the station was small at the bottom, but significantly larger at middle and surface layers. The power law is often used for vertical structural description of flow velocity as follows:

$$\frac{U}{U_0} = (\frac{Z}{Z_0})^{\alpha} \tag{1}$$

Z_0, sea surface measured from the seabed, m
Z, water depth measured from the seabed, m
U_0, current velocity at sea surface, ms^{-1}
U, current velocity at Z depth, ms^{-1}
α, power law exponent

The vertical distribution of measured flow velocity is compared with that of the theoretical curve in Figure 5. The measured data is between α=1/2 and α=1/7 theoretical curves, and does not exactly match one certain curve.

In order to avoid trouble of tidal current flow vertical structure variation, and to visually reflect distribution of tidal current energy in each effective layer(1to 6m above seabed), the average power density of the tidal current energy of each layer during the observation was calculated using the 30min interval tidal current data during one year. The calculation is as follows:

$$P = \frac{0.5\sum_1^n \rho U_i^3}{\text{n}} \tag{2}$$

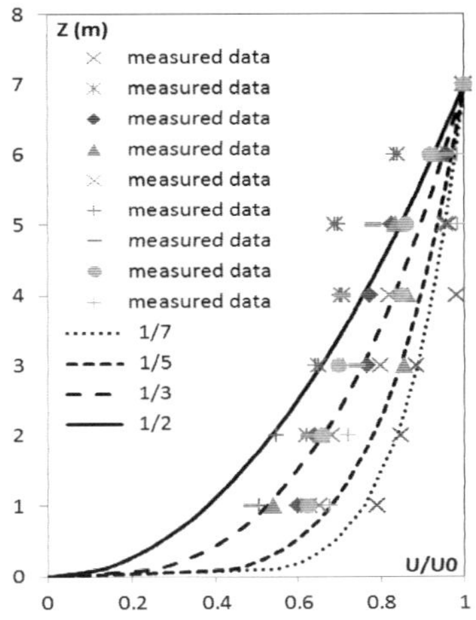

Figure 5. Vertical distribution of velocity

IOP Conf. Series: Earth and Environmental Science **68** (2017) 012017 doi:10.1088/1755-1315/68/1/012017

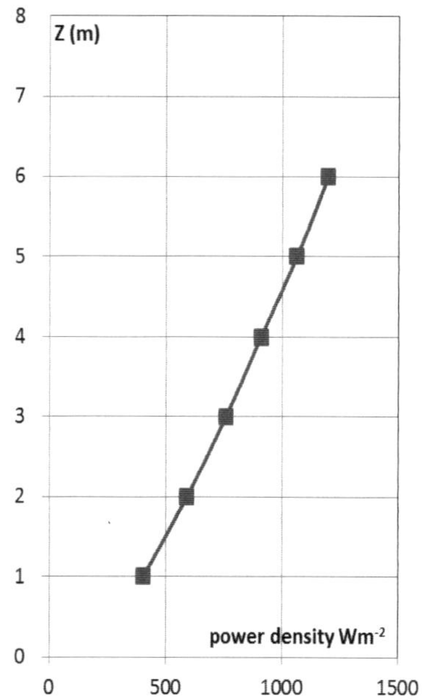

Figure 6. power density for each layer of tidal current energy

P, power density of tidal current energy, average power of tidal current energy in unit area of flowing, Wm^{-2}
ρ, sea water density, 1025 kgm^{-3} is taken
U_i , flow velocity at 30min interval, ms^{-1}
n, number of observations, 17520 in one year
According to the calculation results, vertical distribution of average power density of tidal current energy during observation is shown in Figure 6. Annual average power density of each level is between 401 and 1195 Wm^{-2}. Average power density was lowest at the bottom and highest in surface layer. In shallow water, average power density of tidal current energy decreases obviously to the bottom. The power density drops about 150 Wm^{-2} per meter depth in this site.

In view of the above analysis, more energy can be captured if the tidal current turbine is installed as high as possible until the lowest tide level.

3.3. Direction distribution of tidal current energy
The tidal current turbines being studied or experimented are mostly direction-fixed Horizontal axis turbines[5,7], but tidal current flow changes in a range. So that, the maximum energy that turbine can capture will be limited by angle between the flow direction and turbine orientation. Flow directions at this site are mostly in the range of 240° ± 20° for flood current, and 60 ° ± 20 ° for ebb current. the direction modified power density of tidal current energy was calculated when the turbine is oriented in the range of 220-260 °. The calculation is as follows:

$$P_c = \frac{0.5\sum_1^n \rho \mid U_i \cos\theta \mid^3}{n} \tag{3}$$

P_c, direction modified power density of tidal current energy, Wm^{-2}
ρ, sea water density, 1025kgm^{-3} is taken
U_i, flow velocity at 30min interval, ms^{-1}
θ, angle between flow direction and turbine orientation
n, number of observations, 17520 in one year
The annual average power density of tidal current energy in effective layers (1 to 6m above seabed) is 820 Wm^{-2} regardless the limitation of the angle between flow direction and turbine orientation. In view of limitation of the angle, when the turbine orientation is 299°, direction modified power density

reaches a maximum of 802 Wm^{-2}. When the turbine orientation turns left or right by 10 °, the power density drops 5% to 759 Wm^{-2} (Figure 7).

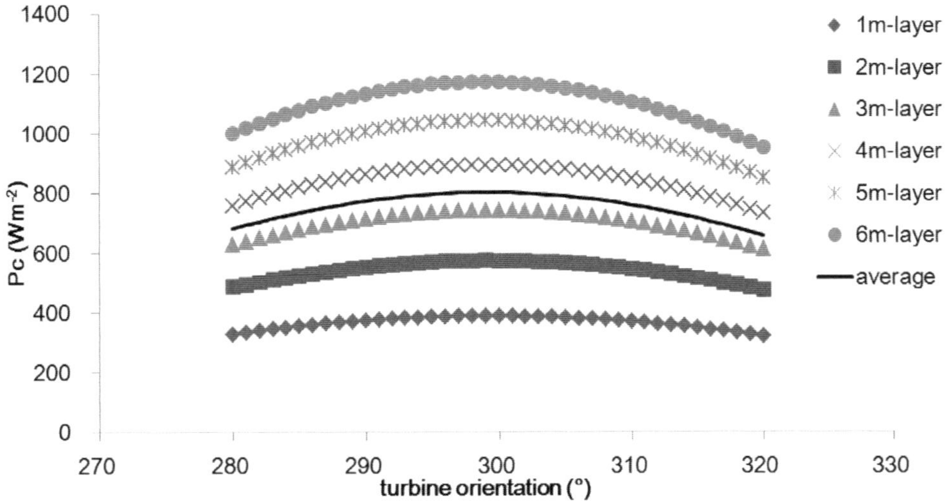

Figure 7. Achievable power for different turbine orientation

3.4. Tidal current energy variation over time

Significant regularity is an important feature of tidal current energy. In semidiurnal tide sea area, it has twice flood time and twice ebb time of high flow speed every day. And there's low speed time between flood and ebb. Flow velocity is also affected by the spring-neap tide cycle. Figure 8 shows a spring-neap cycle in November 2012. In which maximum velocity during spring tide is 2.55 ms^{-1} and the maximum velocity during neap tide is 1.45 ms^{-1}. The daily average power density of current is 1377 Wm^{-2} in spring tide time which is 1.7 times of annual average, and it's 344 Wm^{-2} in neap which is 0.4 times of annual average. Table 3 shows that monthly average power density changes between 648 and 965 Wm^{-2} in different month. The fluctuation range is ± 20% of annual power density, which reminds us to note the unavoidable estimation error using the monthly data in estimating the power density.

3.5. Annual reserve of tidal current energy

The annual average power density calculated by measured date is 820 Wm^{-2}, and direction modified annual power density is 801 Wm^{-2}. However, not all the kinetic energy of tidal current can be captured due to Betz' law and other theory. Efficiency coefficient C_p is developed to calculate the final output of tidal current energy. It is defined as the ratio between captured power and original power. The C_p reported or used by many researchers is usually between 0.3 and 0.5[8]-[10]. In this site, the final annual average power density is 240 Wm^{-2} with a C_p of 0.3. That means the final output will be 59.4MWh by a plant with a 6m-diameter turbine rotor.

Table 3. Monthly average power density from Nov 2012 to Oct 2013(Wm^{-2})

	Nov	Dec	Jan	Feb	Mar	Apr	May	Jun	Jul	Aug	Sep	Oct
P	847	743	721	778	906	823	784	874	1009	869	763	652
P_c	839	736	712	765	892	815	779	848	965	849	757	648

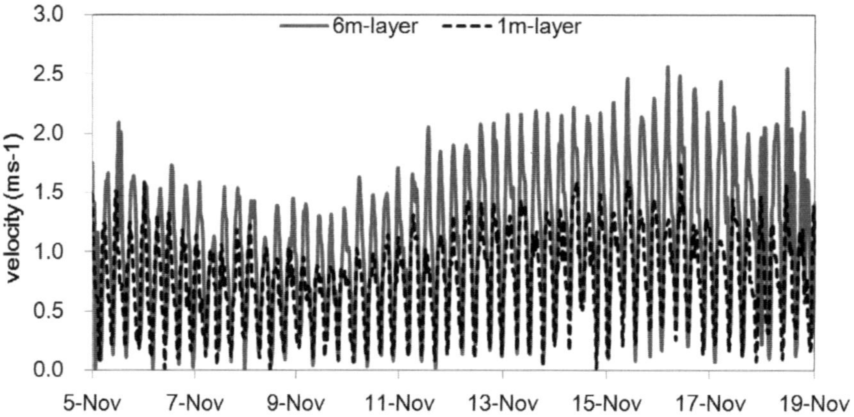

Figure 8. Measured tidal current velocity from Nov 5th to Nov 19th

4. Conclusions

The tidal current is semidiurnal tide in Hangzhou Bay, and it belongs to standing wave. Maximum velocity in effective layers (1 to 6m above seabed) was 2.73 m/s at 6m layer, while maximum depth-average velocity is 2.32 m/s. For one year, flow velocity beyond 0.5m/s occurred in 7010 hours, with guarantee rate of 80%; flow velocity beyond 1m/s occurred in 4406 hours, with guarantee rate of 50%; flow velocity beyond 1.5 m/s occurred in 1113 hours, with guarantee rate of 13%.Monthly average power density changes a lot in different month, and the fluctuation range is about ± 20% of annual power density, which reminds us to note the unavoidable estimation error using the monthly or shorter data in estimating the power density. Average power density of tidal current energy decreases obviously to the bottom. More energy can be captured if the tidal current turbine is installed as high as possible. The direction modified annual average power density is 802 Wm^{-2}. When the turbine turns left or right by 10 °, the power density drops 5% to 759 Wm^{-2}. The final annual average power density with efficiency coefficient C_p of 0.3 is 240 Wm^{-2}. That means the final output of a plant with 6m-diameter turbine rotor will be 59.4MWh in this site.

5. References

[1] Wang Chuan-kun, Lu Wei 2009 Resource analysis method and reserve assessment of ocean energy

[2] Zhang Liang, Li Xinzhong 2013 Tidal current energy update 2013 *Journal of Circuits and Systems* **1**p 53-68

[3] Wu He, Zhao Shiming 2010 Preliminary assessment of tidal current energy on Chengshantou area *Marine Technology* **29** p 98-100.

[4] Lyu Xingang, Qiao Fangli 2010 Numerical Estimation of Marine Tidal Current Energy Resources - Taking Jiaozhou Bay as an Example *Acta Energiae Solaris Sinica* **31** p 127-43

[5] Wang Shi-ming, Ren Wan-chao, Lyu Chao 2016 Review of the tidal current energy device *Marine Science Bulletin* **35** p 601-8

[6] Ramos V, Carballo R, Sanchez M, et al 2014 Tidal stream energy impacts on estuarine circulation *Energy Convers Manage* **80** p 137–49

[7] W-B Chen et al. 2013 Modeling assessment of tidal current energy at Kinmen Island, Taiwan *Renewable Energy* **50** p1073-82

[8] Gorban A N, Gorlov A M, Silantyev V M. 2001 Limits of the turbine efficiency for free fluid flow *Journal of Energy Resources Technology* **123** p 311-7

[9] Lyu Xinguang, Qiao Fangli 2008 Advances in estimation methods for marine tidal current energy resource *Advances in marine science* **26** p 98-108

[10] Bahaj A S, Molland A F, Chaplin, J R, et al 2007 Power and thrust measurements of marine current turbines under hydrodynamic flow conditions in a cavitation tunnel and a towing tank *J. Renew. Energy* **32** p 407–26

Preparation and Performance Evaluation of Biochars from Neem Seed Active Substance Extracted Residues (NSASER)

Jialei Li, Shuwen Lu

Food Processing Institute, Heilongjiang Academy of Agricultural Sciences, Harbin, China

E-mail: lushuwen2016@163.com

Abstract. Neem seed active substance extracted residues (NSASER) is industrial by-product, which is often discarded as a waste. It would lead to a certain degree of harm to the environment. The aim of this study was to prepare the biochars with neem seed active substance extracted residues (NSASER) under the anaerobic pyrolysis conditions. The pyrolysis process was studied with different pyrolysis power (200W, 500W, 800W, 1100W and 1400W), and the performance of the prepared biochars was evaluated. The results showed that the required time was to complete the pyrolysis process that gradually decreased with pyrolysis power increased from 200 to 1400 W, and the final pyrolysis temperature was to complete the pyrolysis process that increased with pyrolysis power increased from 200 to 1400 W. The biochars yield decreased with pyrolysis power increased from 200 to 1400 W, and the biochars yield has the maximum value when the pyrolysis power was of 200W. And the prepared biochars still had some characteristics of the plant cell and kept uniform porous structure, which was beneficial to absorb the small molecule substance. The water content of the prepared biochars was 7.18±0.53, the ash content of the prepared biochars was 5.92±0.31 and the fixed carbon content of the prepared biochars was 81.27±0.89. Compared with the bamboo charcoal, the performance index of the prepared biochars was in according with National Standard of the People's Republic of China GBT26913-2011 of the bamboo charcoal. The prepared biochars had a potential value in application.

1. Introduction

Neem (*Azadirachta indica*) is a native tree, belonging to Meliaceae family which is widely cultivated in tropical and subtropical regions, including Indonesia, Burma and India. Neem has long been used as an herbal medicine in traditional country such as India and Indonesia [1]. It has long been used in traditional country so as to prevent and treat various diseases such as malaria and so on. In recent years, many modern scientific and medical studies have demonstrated that it possessed a wide spectrum of biological activities such as antibacterial and antifungal [2]. Validation studies conducted confirm that neem has been rich in much secondary metabolites such as terpenoids. Various parts of neem could be used as an herbal medicine, including leaves, root bark, stem bark and seed [3]. Among them, neem seed is most widely used as pesticides and medicine [4]. It has been proved that azadirachtin and neem seed oil were the main active substance of the neem seed, which could prevent and treat various diseases [5]. Azadirachtin is one of the best natural plant sources for the development of biological pesticides [6]. Also, Azadirachtin is widely used in medicine such as malaria, fever and fertility. Neem seed oils have a good antifertility effect, and have not yet found its toxic and side effects. It is a kind of natural plant medicine with great research and development value [7]. Usually, the extraction of azadirachtin was carried out using solvent extraction method. In the extraction of azadirachtin, neem seed oil was also often isolated [8]. However, these neem seed active substance extracted residues (NSASER) with small amount of organic reagents are often treated simply and then thrown away. It leads to a certain degree of harm to the environment.

Biochars is a kind of carbonaceous materials by incomplete combustion of biomass materials under anaerobic or oxygen-limited conditions [9]. Biochars has the features of high carbon content, high stability, abundant surface functional groups, multiple porosity, high specific surface area and cation exchange capacity and so on, which could be used as the adsorbents for heavy metals and organic pollutants in soil and water. Also, it has the potential to fix heavy metals and organic pollutants in the soil [10], [11]. Therefore, biochars could be widely applied to many aspects such as fixing carbon, reducing discharge, retarding greenhouse gas emission, charring field, improving soil, promoting crop growth and remediation of environmental pollution [12], [13]. Biochars could be produced from much biological resources, including forest residues, agricultural waste, urban waste, yard waste and so on [14]. Neem seed active substance extracted residues (NSASER) was belonging to agricultural waste, and so it could be regarded as a raw material to be used to prepare biochars.

The aim of this study was to prepare the biochars with neem seed active substance extracted residues (NSASER) under the anaerobic pyrolysis conditions. The pyrolysis process was studied with different pyrolysis power. Then, the performance of the prepared biochars was evaluated. It makes neem seed active substance extracted residues (NSASER) be effectively utilized.

2. Materials and methods

2.1. Material

The neem seed active substance extracted residues were collected after the azadirachtin and neem seed oils were extracted.

2.2. Preparation of biochars

Preparation of biochars was usually used by direct carburation [15], and it was described by the following processes, as shown in Fig. 1. Firstly, the neem seed active substance extracted residues (NSASER) was dried by hot-air drying at 40℃. Then, the neem seed active substance extracted residues (NSASER) was smashed and then sieved through 10-mesh before use. 200 g of the neem seed active substance extracted residues (NSASER) powder was weighed precisely, and put into the self-made anaerobic pyrolysis device, as shown in Fig. 2. Subsequently, the anaerobic pyrolysis experiments were performed with a certain pyrolysis power. After that the biochars were obtained.

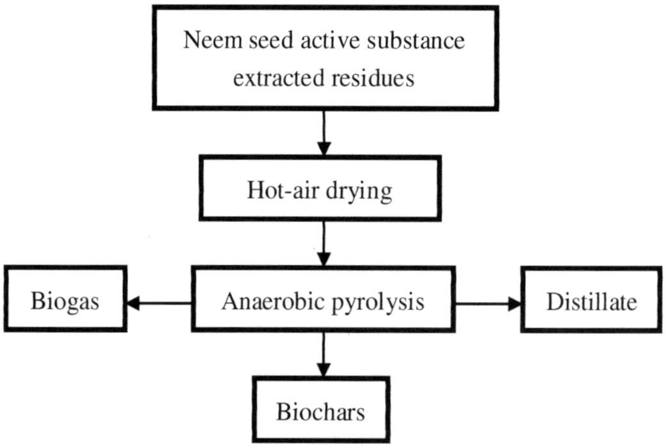

Figure 1. Flowchart of anaerobic pyrolysis

2.3. Analysis of anaerobic pyrolysis process

Effect of different pyrolysis power (200W, 500W, 800W, 1100W and 1400W) on the process was studied. The value of the change in time and temperature was record with the self-made anaerobic pyrolysis device. Also, Effect of different pyrolysis power (200W, 500W, 800W, 1100W and 1400W) on the biochars yield was studied. Weighing method was applied here to measure the biochars yield. The biochars yield (%) was calculated as following Eq. (1):

$$A = \frac{A_1}{A_0} \times 100\% \tag{1}$$

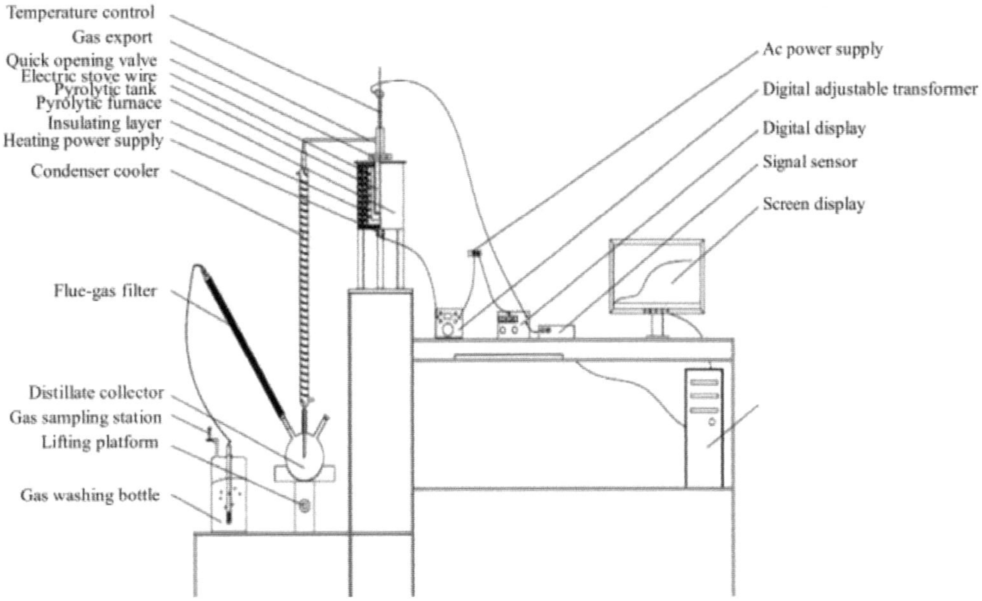

Figure 2. Anaerobic pyrolysis device

Where A is the biochars yield, A_1 is the obtained biochars weight, and A_0 is the raw material weigh.

2.4. Scanning electron microscope

The prepared biochars were coated with gold and examined with a scanning electron microscope system under high vacuum condition at an accelerating voltage of 15.0 kV, as well as image magnifications 2000 ×.

2.5. Determination of performance index of biochars

2.5.1. Determination of water content

Determination of water content was described as National Standard of the People's Republic of China GBT26913-2011 [16]. The prepared biochars was smashed and then sieved through 18-mesh before use. 10 mL of weighing bottle was put into the drying oven, and dried at 105℃ to constant weight. Then, 10 mL of weighing bottle was weighed after cooling 30 min. 1 g of prepared biochars was weighed, and placed in the above mentioned treated 10 mL of weighing bottle, which makes the prepared biochars distribute homogeneously. Then, it was put into the drying oven, and dried at 105℃ for 4 h. After that it was taken out and put into the dryer for 30 min. After cooling, 10 mL of weighing bottle with the prepared biochars was weighed. The water content (%) was calculated as following Eq. (2):

$$B = \frac{B_0 - B_1}{B_0 - B_2} \times 100\% \tag{2}$$

IOP Conf. Series: Earth and Environmental Science **68** (2017) 012018 doi:10.1088/1755-1315/68/1/012018

Where B is the water content of the biochars, B_0 is 10 mL of weighing bottle with the biochars weight before drying, B_1 is 10 mL of weighing bottle with the prepared biochars weight after drying, and B_2 is 10 mL of weighing bottle weight.

2.5.2. Determination of ash content

Determination of ash content was described as National Standard of the People's Republic of China GBT26913-2011 [16]. The prepared biochars was smashed and then sieved through 55-mesh before use. 30 mL of porcelain crucible was put into the high temperature electric furnace, and calcined at 800℃ to constant weight. Then, 30 mL of porcelain crucible was weighed after cooling 30 min. 1 g of dried prepared biochars was weighed, and placed in the above mentioned treated 30 mL of porcelain crucible. Then, it was put into the high temperature electric furnace below 300℃, then slowly heated to 500℃ for 30 min. After that it was slowly heated to 800℃ for 4h. After cooling, 30 mL of porcelain crucible with the prepared biochars was weighed. The ash content (%) was calculated as following Eq. (3):

$$C = \frac{C_2 - C_1}{C_0} \times 100\% \tag{3}$$

Where C is the ash content of the biochars, C_0 is the biochars weight, C_1 is 30 mL of porcelain crucible weight, and C_2 is 30 mL of porcelain crucible with the ash weight.

2.5.3. Determination of volatile content

Determination of volatile content was described as National Standard of the People's Republic of China GBT26913-2011 [16]. The prepared biochars was smashed and then sieved through 55-mesh before use. 1 g of dried prepared biochars was weighed, and placed in the above mentioned treated 30 mL of porcelain crucible, which makes the prepared biochars distribute homogeneously. Then, it was quickly put into the high temperature electric furnace with 850℃, and heated for 7 min. 30 mL of porcelain crucible should be positioned above or below the galvanic measuring point. After that it was taken out and cooled for 5 min in the air. Then, it was put into the dryer to be cooled to room temperature, and then weighed. The volatile content (%) was calculated as following Eq. (4):

$$D = \frac{D_1 - D_2}{D_1} \times 100\% \tag{4}$$

Where D is the volatile content of the biochars, D_1 is the biochars weight, and D_2 is the biochars weight after heating.

2.5.4. Determination of fixed carbon content

Determination of fixed carbon content was described as National Standard of the People's Republic of China GBT26913-2011 [16]. The fixed carbon content (%) was calculated as following Eq. (5):

$$E = 100 - (C+D) \tag{5}$$

Where E is the fixed carbon content of the biochars, C is the ash content of the biochars, and D is the volatile content of the biochars.

2.6. Statistical analysis

IOP Conf. Series: Earth and Environmental Science **68** (2017) 012018 doi:10.1088/1755-1315/68/1/012018

The obtained data were presented as mean ± standard deviation (SD), and all of the measurements were made in triplicate. All obtained data were analyzed using the OriginPro version 8.0 software.

3. Results and discussion

3.1. Effect of pyrolysis power on the process

For investigating the effect of pyrolysis power on the process, it was to set pyrolysis power at 200W, 500W, 800W, 1100W and 1400W, respectively. During the anaerobic pyrolysis process, the value of the change in time and temperature was record with the self-made anaerobic pyrolysis device. Then, the pyrolysis curves were plotted, as shown in Fig. 3. As seen from Fig. 3, the required time was to complete the pyrolysis process that gradually decreased with pyrolysis power increased from 200 to 1400 W, and the final pyrolysis temperature was to complete the pyrolysis process that increased with pyrolysis power increased from 200 to 1400 W. Under different pyrolysis power conditions, there was always a period of temperatures rising slowly at the beginning of the anaerobic pyrolysis process. That is because the moisture in the neem seed active substance extracted residues (NSASER) has absorbed heat in the evaporation process.

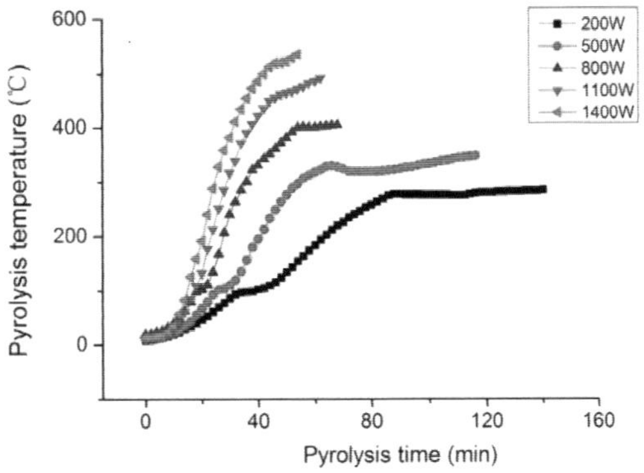

Figure 3. Pyrolysis curve of different pyrolysis power

3.2. Effect of pyrolysis power on biochars yield

For investigating the effect of pyrolysis power on biochars yield, it was to set pyrolysis power at 200W, 500W, 800W, 1100W and 1400W, respectively. The effect of pyrolysis power on biochars yield was as shown in Fig. 4. The biochars yield decreased with pyrolysis power increased from 200 to 1400 W. That is because there are three kinds of products during the anaerobic pyrolysis process, including the biochars, the biogas and the distillate (biomass tar and biomass vinegar). The biogas and the distillate (biomass tar and biomass vinegar) increased with pyrolysis power increased from 200 to 1400 W, which makes the proportion of biochars in all products reduced. The biochars yield has the maximum value when the pyrolysis power was of 200W.

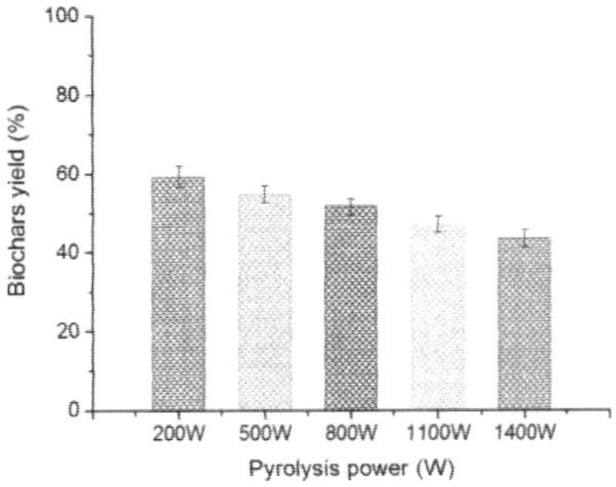

Figure 4. Effect of different pyrolysis power on biochars yield

3.3. Scanning electron microscope of the prepared biochars

The biochars was prepared by pyrolysis power of 200W. Then, the prepared biochars were coated with gold and examined with a scanning electron microscope system. The scanning electron microscope images provide a visual evidence of the structure of the biochars. As shown in Fig. 5, the prepared biochars still had some characteristics of the plant cell and kept uniform porous structure. It was beneficial to absorb the small molecule substance.

Figure 5. Scanning electron microscope images of the prepared biochars by pyrolysis power of 200W

3.4. Performance evaluation of the prepared biochars

The water content, the ash content and the fixed carbon content were important performance index to evaluate biochars. For evaluating the performance of the prepared biochars by pyrolysis power of 200W, the water content, the ash content and the fixed carbon content were determined by the method described as National Standard of the People's Republic of China GBT26913-2011 [16]. As a result, the water content of the prepared biochars was 7.18±0.53, the ash content of the prepared biochars was 5.92±0.31 and the fixed carbon content of the prepared biochars was 81.27±0.89, as shown in Table 1. Among all the biochars, the bamboo charcoal is one of the most widely used. As seen from Table 1, compared with the bamboo charcoal, the performance index of the prepared biochars was in according with National Standard of the People's Republic of China of the bamboo charcoal. The prepared biochars had a potential value in application.

Table 1. Performance index of bamboo charcoal* and prepared biochars.

Performance index	Bamboo charcoal*		Prepared biochars
	First grade product	Qualified product	
Water content (%)	≤9.00	≤12.00	7.18±0.53
Ash content (%)	≤4.50	≤6.50	5.92±0.31
Fixed carbon content (%)	≥85.00	≥75.00	81.27±0.89

*National Standard of the People's Republic of China GBT26913-2011

4. Conclusion

Neem seed active substance extracted residues (NSASER) is industrial by-product, which is often discarded as a waste. It would lead to a certain degree of harm to the environment. The aim of this study was to prepare the biochars with neem seed active substance extracted residues (NSASER) under the anaerobic pyrolysis conditions. The pyrolysis process was studied with different pyrolysis power (200W, 500W, 800W, 1100W and 1400W), and the performance of the prepared biochars was evaluated. The results showed that the required time was to complete the pyrolysis process that gradually decreased with pyrolysis power increased from 200 to 1400 W, and the final pyrolysis temperature was to complete the pyrolysis process that increased with pyrolysis power increased from 200 to 1400 W. Under different pyrolysis power conditions, there was always a period of temperatures rising slowly at the beginning of the anaerobic pyrolysis process. The biochars yield decreased with pyrolysis power increased from 200 to 1400 W, and the biochars yield has the maximum value when the pyrolysis power was of 200W. And the prepared biochars still had some characteristics of the plant cell and kept uniform porous structure, which was beneficial to absorb the small molecule substance. The water content of the prepared biochars was 7.18±0.53, the ash content of the prepared biochars was 5.92±0.31 and the fixed carbon content of the prepared biochars was 81.27±0.89. Compared with the bamboo charcoal, the performance index of the prepared biochars was in according with National Standard of the People's Republic of China of the bamboo charcoal. The prepared biochars had a potential value in application.

5. Acknowledgments
This work was financially supported by the Postdoctoral Foundation of Heilongjiang Province (LBH–Z12243), Harbin Special Fund for Scientific and Technological Innovation Talents (2016RAQXJ196) and Doctoral Fund of Heilongjiang Academy of Agricultural Sciences (201507-44).

6. References
[1] Benelli G, Bedini S, Cosci F, Toniolo C and Conti B 2015 *Parasitology research.* **114**(1) 227，36
[2] Srinivasa K, Jagadeesh K and Revankar S P 2014 *Journal of Medical and Health Sciences.* **3**(1) 93-96
[3] Bharali R K and Bhattacharyya K G 2015 *Journal of Environmental Chemical Engineering.* **3**(2) 662-669
[4] Opoku A, Chaves B and De Neve S 2014 *Biological agriculture & horticulture.* **30**(3) 145-152
[5] Betiku E, Omilakin O R, Ajala S O, Okeleye A A, Taiwo A E and Solomon B O 2014 *Energy.* **72** 266-273
[6] Abedi Z, Saber M, Gharekhani G, Mehrvar A and Kamita S G 2014 *Journal of economic entomology.* **107**(2) 638-645
[7] Balaji G and Cheralathan M 2014 *Journal of the Energy Institute.* **87**(3) 188-195
[8] De Paula J A M, Brito L F, Caetano K L F N, De Morais Rodrigues M C, Borges L L and Da Conceição E C 2016 *Talanta.* **149** 77-84
[9] Azargohar R, Nanda S, Kozinski J A, Dalai A K and Sutarto R 2014 *Fuel.* **125** 90-100
[10] Kim K H, Kim T S, Lee S M, Choi D, Yeo H, Choi I G and Choi J W 2013 *Renewable Energy.* **50** 188-195
[11] Huang D, Wang Y, Zhang C, Zeng G, Lai C, Wan J, Qin L and Zeng Y 2016 *RSC Advances.* **6**(77) 73186-73196
[12] Purakayastha T J, Das K C, Gaskin J, Harris K, Smith J L and Kumari S 2016 *Soil and Tillage Research.* **155** 107-115
[13] Ahmad M, Rajapaksha A U, Lim J E, Zhang M, Bolan N, Mohan D, Vithanage M, Lee S S and Ok Y S 2014 *Chemosphere.* **99** 19-33
[14] Shi Y, Ge Y, Chang J, Shao H and Tang Y 2013 *Renewable and Sustainable Energy Reviews.* **22** 432-437
[15] Kim J H, Ok Y S, Choi G H and Park B J 2015 *Chemosphere.* **134** 435-437
[16] National Standard of the People's Republic of China GB/T 29403-2012, "Bamboo charcoal".

Critical Review of Removal of Nano Materials in Waste Streams

Solomon W. Leung[1], Bradley Williams[1], Karl De Jesus[2], and James C.K. Lai[3]

[1]Civil and Environmental Engineering Department at Idaho State University, Idaho, USA
[2]Chemistry at Idaho State University.
[3]Pharmacology & Toxicology at Idaho State University College of Pharmacy.

Abstract. Industrial applications of nanomaterials (NMs) are rising drastically in recent years and the commercial value of these materials can reach over $100 billion in 5 years. Major effort in nano research has been devoted to the utilities of the materials, only minimum effort has been directed to the disposal, reuse, and recycle of these new forms of materials. Due to their unique sizes and sharps, nanomaterials possess unique characteristics and toxicity that are not expected from their counterparts in meso/micro forms. At the present time, there are no regulations governing the handling and disposals of NMs, but recent research demonstrated that NMs are more hazardous than we realize. A main reason why less caution is being exercised by the general public regarding NMs is that the measurement and quantitation of NMs are difficult, which lead to difficulties in monitoring, thus regulation. This article critically reviewed over the issues stemming from the development of NMs, especially the challenges of measurement and disposal of these materials in landfills.

1. Introduction

Industrial applications of nanomaterials (NMs) in industries in recent years are expected to reach over $100 billion in the global nanotechnology market in 2021 [1]. Research in NMs and their fate and sustainability in waste streams, eventually the environment, has highlighted some major concerns. This paper will discuss the literature search, current works, findings, and some relevant results that are associated with the generation and removal of NMs.

To begin this process of understanding the current leading edge knowledge and active work in the subject of nanomaterial (NM) as it becomes waste, an extensive literature search was conducted. Hundreds of articles and studies have been published on many different aspects regarding NM, ranging from transport, toxicology, and environmental accumulation, to uptake including in animals and plants. Currently NMs are used in industries worth multibillion dollars and the trend is in the upswing, however, limited amount of commercial effort has been directed to capture and recovery of these new forms of materials.

To better understand the scope of NM impact to human health and the environment, and the need for recovery processes, this paper will highlight the levels of NMs being generated, bioavailability, toxicology, measurement methods, and possible recovery solutions. Most important of all, since NMs are being generated for commercial use, these MNs ultimately end up entering a typical municipal wastewater treatment plant (WWTP) or landfill. Once NMs enter the WWTP they are either captured in the waste solids or are discharged in the effluent. The persistent NMs stay in nano form either in the liquid or solids and then are released to the environment. Waste solids are usually transported to a landfill and are incinerated, or used as mulch or fertilizer onto an open field. Liquid as effluent is therefore released back into surface waters. This general practice of NM management can then lead to NMs unintentionally reentry into the environment and possibly the food chain that is not a desirable solution.

2. Nanomaterial Industrial Waste Discharge Amounts (2010 data)

In looking at nanomaterial in waste streams and possibilities for recovery, one of the first items that need to be understood is the amount of industrial nanomaterial being produced and released to waste. A study conducted by. Keller et al. [2], based on material flow modeling discharged into waste streams, showed the top 10 (by metric tons/year) most abundant engineered nanomaterials (ENMs) from 2010 global data: the study calculated how much ENM was produced and how much would then be discharged as shown in Table 1. The amount discharged to waste was based on presumptive material flow and then back calculated by knowing the reported amount produced rather than the actual measurement of the true ENM concentration in a waste stream.

Table 1: Top 10 ENMs AND DISCHARGES (YEAR/METRIC TON) IN 2010.

	Engineered nano material	Discharge
1	Silica (SiO_2)	95,000
2	Titania (TiO_2)	88,000
3	Iron oxides	41,900
4	Alumina (Al_2O_3)	34,900
5	Zinc oxides	34,100
6	Nano-clays	10,400
7	CeO_2	10,000
8	Carbon nano tube (CNT	3,276
9	Cu and Cu oxides	497
10	Ag	424

The next important factor to take into account when evaluating nanomaterial in waste streams is how long will each ENM remain in nanoparticle form. This means that ENMs that are soluble and would quickly go into solution immediately are no longer in the nanoparticle form; also ENMs that aggregate are becoming larger than the original nanoparticles would be classified as bulk material. For examples: Iron, Zinc, and Cu oxides would likely not last very long in the nanoparticle form as they are more soluble. On the other hand, TiO2 is one of the least soluble among ENMs and would likely stay in the nanoparticle form for much longer. SiO2 would also be very similar to TiO2 with a low solubility.
The top 3 industries that account for 42% of the total ENMs release are: 1) Coatings, 2) Paint and Pigments, and 3) Cosmetics. Nearly all the ENMs are released into landfills and/or soils, with the remaining in wastewater streams [2]. Some of these releases to the landfills come originally from municipal wastewater and then are collected as sludge of the treatment plant and eventually end up in landfills.

3. Case Studies: True Measurements of NM

Two case studies are cited in this study to elucidate some of the difficulties that are encountered in NM measurements. The first case study was conducted in Rome, Italy. This study measured concentration of nanocontaminants (NCs) that entered the municipal wastewater treatment plants (WWTPs) and then the concentrations in the effluent. In addition bulk samples were taken in the receiving waters to measure the concentrations of the same NCs [3]. It was found that three NCs had high pass through and posted risk in the environment. They are:

1) Carbamazeoine with 0.24 μg/l;
2) Estrone with 0.56 μg/l;
3) Gemfibrozil with 0.11 μg/l.

It should be noted that estrone showed basically no removal in the WWTP. It is one of several natural estrogens that is carcinogenic to females and can cause erectile dysfunction to men.

Next was a study of Darwin Harbor Estuary in northern Australia [4]. This study took samples throughout water bodies adjacent to a heavily populated municipality. The location was selected due to a high rate of population growth in the area. In this study, carbamazepine, estrogen, and testosterone concentrations were found in concentrations up to 0.5 μg/l, 0.33 μg/l, and 6.29 ng/l, respectively.

Even though the above studies did not look for ENMs specifically, there were significantly large amounts of hormones and pharmaceutical wastes found and with increasing concentrations near populated areas. These studies indicated that more hazardous wastes are discharged into our surface waters system that we are not aware of at metropolitan areas.

On a good note, it has been shown that wetland can be used to treat NCs to undetectable levels. A study from the Czech Republic showed that wetlands removed all incoming hormones to below detection limits [5]. The incoming levels of wastes that had 55 ng/l of estrone and 11 ng/l of testosterone initially were all removed by constructed wetlands.

4. Modeling and Transport of NM

There are two aspects in looking at modeling of NMs: First is the overall mass balance of how much NMs are produced, and then the eventual final deposal of the NMs. The latter is the more detailed studies of targeted NMs and their movements through waste streams and the resultant environmental impacts. A study of global life cycle release of NMs showed excessive amount of releases from cosmetics, textiles, paint and coatings, and filtration industries [2]. Majority of these wasted NMs are to be deposited in landfills and soil, with quite a large portion of such remains in waste water that is discharged into the surface waters.

There were two case studies that modeled NMs passing through WWTP and were field sprayed onto the soils. Both studies showed that NM concentrations were increasing [6]-[7]. But when these models were compared to actual measurements the modeled values were still less than the measurements, which suggests that the models are conservative and the emissions of NMs may be greater than originally expected.

5. Bioavailability of NM

A number of different studies have shown that nanoparticles are taken up by plant life either from the soil or water column. These particles maintain their nano size and therefore would be also suspected to have toxic effects. One study showed application of carbon nanomaterials to rice plants, the nanomaterial have been observed existing in various tissues of the plants [8]-[10]. This suggests that as environmental levels on ENMs increase, there is a higher risk of the nanomaterial to reenter the food chain. Another study showed that ZnO nanoparticles had a phytotoxic impact on Ryegrass [11].

The key point of bioavailability is that either ENMs are released to the environment via WWTP effluent as suspending solids, or are trapped in waste solids that are used as mulch or fertilizer, there is a clear evidence that NMs will be taken up by plant life, and may eventually be part of the human food-chain. In addition, how long and how many cycles these NMs remain in nano sizes and their toxicity are not known at this time.

6. Toxicity of NM

Although NMs are recent discovery, considerable amount of research has been conducted on the topic. Majority of the results were verified by laboratory experiments. NMs have shown a unique toxicological effect that is not previously expected. For many cases when one would consider the toxicity of a bulk substance that appears to be benign, the size and shape of the same substance in nano form would now make it toxic to living cells. A basic explanation is that with particles that are so small that they can pass through cell walls that they will likely have a different impact on that cell than the same bulk substance that cannot pass through the cell wall.

A second part to toxicity is the inclusion of hormones and pharmaceuticals. In the nano contaminant (NC) form many hormones and pharmaceuticals will reenter the body and cause developmental and adverse impacts on human health. In the case of ENMs, they are designed to maintain the size and shape for an extended life depending on the application. But in the case of hormones and pharmaceuticals they are not intended to still be functioning after their initial use in the body. Their persistence in the environment is causing unexpected health consequence.

With all the variety of ENMs, TiO2 seems to have a large number of related studies. This is likely due to TiO2 being one of the more stable ENMs and is used in many industrial and domestic applications, thus exists in high abundancy. Taking this in mind most of the following discussion is in reference to TiO2 [12]-[13].

TiO_2 is classified as a Group 2B carcinogen, with nano-$Ti0_2$ having stronger catalytic activity than its non-nano counterpart; the nano particle has gained much more toxicological attention in recent studies [14]. Exposure to nano-$Ti0_2$ can induce inflammation, cytotoxicity, genotoxicity, and phototoxicity depending on the cells and form of exposure. [15]-[16]. Nano-TiO_2 particles have many different pathways to enter the human body: 1) Injection in to the blood; 2) Inhalation in to the respiratory tract; 3) Ingestion into the gastrointestinal tract; and 4) Dermal through the skin. It was shown that nano-TiO_2 can penetrate the skin causing damage to different organs [17]. Also nano-TiO_2 (anatase) is phototoxic to skin due to its photo-reactivity when exposed to UV light [18]. This is a concern because one of the major uses of nano-TiO_2 is in sun screen. TiO_2 has been shown in many different studies to have the ability to be transported to many different organs along with its ability to crossing over the blood brain barrier. Liver and kidney were shown to easily accumulate nano-TiO_2. Toxic effects have been observed in tests on specific cells in rats or mice, showing impairment and/or cell death. It has been shown that exposure of $TiO_{2,}$ specifically to brain cells, can cause damage leading to neurodegenerative diseases [19].

7. Challenges in Accurate Measurements of NM

As the toxicity of NMs was unexpected, also as difficult is the current methods to accurately measure the sizes and concentrations of NMs in environmental samples on a broad scale. Acquiring real concentrations of NMs are difficult at best when the NMs being measured are actually known to be existing, let alone the presence of NMs in a sample are unknown. When looking at NMs there are many characteristics that need to be evaluated, such as the average particle size, shape, elemental composition, and mass concentration. Size and shape can be very critical in relating to toxicity [20]. Especially the size and shape of NMs, these characteristics are dependent of the bulk properties of the environmental media, such as pH, chemical affinity between the NMs and the solution they are in, and the ability of polymerization by the NMs. For example, if a NM has the size and shape that allows it to cross the blood brain barrier, then there is a very high risk to human health regardless of the material composition. NM is very difficult to differentiate between ENM and its natural counterparts in an unknown waste sample. As imaging and material science are advancing, better characterization and measurement techniques are definitely in demand.

8. Measurement Methods

There have been extensive studies evaluating specific NMs. In doing so, each study has to have a method for analyzing the concentration, chemical makeup and particle size of the nanomaterial. Each method is a little different depending on the sampling medium. For liquid samples such as WWTP effluents, large volume samples need to be collected and concentrated using a process such as rotary-evaporation and dialysis then freeze drying [21]. Then ICP (inductively coupled plasma), EDX (energy-dispersive X-ray) and TEM (transmission electron microscopy) analysis are performed on the samples.

Just as important in the analysis is the separation of different colloids. This is done by acid reduction depending on the types of colloid of interest such as organic or inorganic. This works on ENM such as TiO_2 because it mostly remains in its original form. Other ENMs that may dissolve or react and transfer into other forms must be analyzed by different methods.

9. Removal mechanisms

By the nature of NM, being of such a small particle size, normal particle settling dynamics are not relevant. Next important characteristic to know about persistent NMs is that some of these particles also resist aggregation due to the surface chemistry and surface charge [22]. Nanoparticles that cannot be aggregated will then subsequently end up in the WWTP effluent. Nanoparticles that can be aggregated will be collected in the WWTP solids [23]-[24].

In viewing ENMs for removal, we should first consider the most environmentally persistent NMs. TiO_2 is one of the most environmentally persistent NMs and is a good candidate as an ideal surrogate for removal and recovery study. To begin understanding possible ENM removal solutions, it is important to highlight the key examples that showed some kind of characteristics that supported aggregation. In a study that concentrated on the aggregation behavior of nanoparticles in WWTP, showed that no aggregation was observed for a variety of nanoparticles in wastewater effluent [25]-[26]. But then when the test was repeated with wastewater containing dissolved organic matter (DOM), aggregation was observed. This demonstrated how the change of surface chemistry induced by organic matters may affect the state of NMs.

Furthermore, vast majority of nanoparticles that are captured in the WWTP processes utilize organic processes such as activated sludge or trickling filter. This identifies a possible mechanism for changing the surface chemistry to allow for aggregation. Also shown in studies is that ENM will be adsorbed to biomass [12].

The next step to be considered is recovery. This is because even though the NPs are adsorbed to biomass, many will still maintain their nano size or be very loosely aggregated, meaning even though NPs are incorporated onto the bulk solid they maintain the characteristics of NM. Often, when NPs are captured in a solid, the clean-up/disposal may have been considered completed. But with NPs, they maintain their nano-characteristics and are just temporarily "stored" in the bulk solid, a good example is sludge in WWTP. A general practice for sludge management is to spread the sludge in an open field as final disposal. If the NPs captured in the sludge retain the nano-properties, field application would unintentionally allow the NPs return to the environment. Thus, it becomes clear that immediate NM recovery is necessary. In doing the recovery to make the ENMs viable again for commercial use they have to be separated from the bulk mass and the surface properties restored. For example, the ENMs would need to be separated into an identifiable specific nanoparticle (e.g., TiO_2) with a reasonable level of purity to enable recycling. Further research is needed to encourage this recycle/reuse initiative.

10. Possible Removal Techniques

Taking the characteristics that are favorable for separation and isolation, techniques for actually removing and capturing nanoparticles needs to be developed. One such method is the modification of procedures during incineration of the WWTP solid effluent, which is already a normal practice with many plants. Taking this process and adding scrubbing technique of the exhaust gases would likely result in reasonable nanoparticle capturing rate. One possibility may be spraying the exhaust gases with a solution of calcium carbonate that is widely used in the industry. At this point the nanoparticles would be suspended in solution that are mixed with other particles (such as carbon ash and other metals) and require further separation and drying.

For treatment and removal, in applications that do not go through a WWTP or that are in the liquid effluent of the WWTP, an organic based coagulation and settling type process would be needed to pull the nanoparticles from suspension. It was shown in a number of studies that nanoparticle removal rate was greatly increased by the presence of organics [25].

In all, the removal and capture of nanoparticles will take a series of steps incorporating two main processes: The first step is changing the surface chemistry to allow aggregation and then settling of the NPs. The next step is drying the solids and then remove the intermediate material that was used to promote aggregation in the first step. A process to get reusable original form nanoparticles may be challenging, experimental trials are needed to ensure a practical and working process. At this time, there is no one known universal removal/recycle technique for these multifaceted properties of NPs.

11. Summary

Nanomaterials have been proven to be very useful in many applications and are being produced at ever increasing rates. At the same time, NMs possess unexpected toxicity and risk to human health due to size and shape that are not typically observed in their bulk material counterparts. Furthermore, NMs are not being captured or recovered by the vast majority of waste treatment methods and they are being released into the environment at an alarming rate. Whether the NMs are in the effluent or the waste solids of the WWTPs, some of them remain their unique characteristics that they can be a health concern if they are released into the environment and be part of the food chain. The next critical point is that measurement of unknown waste or environmental samples for NMs is inaccurate and often near impossible. Currently, there are no regulations governing NM disposal that is becoming a public health concern.

In summary, we could be just seeing the tip of the iceberg of a possible significant environmental impact on human health that would be reminiscent of such events seen during the industrial revolution. The alarming issues with NMs are the difficulties of detection and measurement, then the realization of hazard, in addition to the disposal of the wastes. The fact that the secondary environmental concentrations are non-existing because these materials cannot be practically sampled. Disposal of NMs is an imminent issue that deserves more research and regulatory attention.

12. References

[1] McWilliams, Andrew, "The maturing nanotechnology market: Products and applications," BCC Research LLC, Wellesley, MA, 2016.

[2] Arturo A. KELLER, SUZANNE MCFERRAN, ANASTASIYA LAZAREVA, AND SANGWON SUH, "GLOBAL LIFE CYCLE RELEASES OF ENGINEERED NANOMATERIALS," *JOURNAL OF NANOPARTICLE RESEARCH*, VOL. 15, NO. 6, PP. 1-17, 2013.

[3] Luisa PATROLECCO, SILVIO CAPRI, AND NICOLETTA ABEMOLLO;, "OCCURANCE OF SELECTED PHARMACEUTICALS IN THE PRINCIPAL SEWAGE TREATMENT PLANTS IN ROME (ITALY) AND IN THE RECEVING SURFACE WATERS," *ENVIRON SCI PULLUT RES*, PP. 5864-5876, 2015.

[4] Veronica A. FRENCH, SUSAN CODI KING, ANU KUMAR, GRANT NORTHCOTT, KEITH McGuinness, AND DAVID PARRY, "CHARACTERISATION OF MICROCONTAMINANTS IN DARWIN HARBOUR, A TROPICAL ESTUARY OF NORTHERN AUSTRALIA UNDERGOING RAPID DEVELOPMENT," *SCIENCE OF THE TOTAL ENVIRONMENT*, PP. 639-647, 2015.

[5] Jan Vymazal, Tereza Brezinova, and Milan Kozeluh, "Occurrence and removal of estrogen, progesterone and testosterone in three constructed wetlands treating municipal sewage in the Czech Republic," *Science of the Total Environment*, pp. 625-631, 2015.

[6] Tian Yin Sun, Gulliver Conroy, Erica Donner, Konrad Hungerbuehler, Enzo Lombi, and Bernd Nowack, "Probabilistic modelling of engineered nanomaterial emissions to the environment: a spatio-temporal approach," *Environmental Science-Nano*, vol. 2, pp. 340-351, 2015.

[7] Christine Ogilvie Hendren, Appala R. Badireddy, Elizabeth Casman, and Mark R. Wiesner, "Modeling nanomaterial fate in wastewater treatment: Monte Carlo simulation of silver nanoparticles (nano-Ag)," *Science of the Total Environment*, vol. 449, pp. 418-425, 2013.

[8] S. Lin, J. Reppert, Q. Hu, J. Hudson, M. Reid, T. Ratnikova, A. Rao, H. Luo, and P. Ke, "Uptake, translocation, and transmission of carbon nanomaterials in rice plants," *Small*, vol. 5, no. 10, pp. 1128-1132, May 2009.

[9] H. Zhu, J. Han, J. Xiao, and Y. Jin, "Uptake, translocation, and accumulation of manufactured iron oxide nanoparticles by pumpkin plants," *Journal of Environmental Monitoring*, vol. 10, no. 6, pp. 713-717, 2008.

[10] W. Lee, Y. An, H. Yoon, and H. Kweon, "Toxicity and bioavailability of copper nanoparticles to the terrestrial plants mung bean (Phaseolus radiatus) and wheat (Triticum aestivum): Plant agar test for water-insoluble nanoparticles," *Environmental Toxicology and Chemestry*, vol. 27, no. 9, pp. 1915-1921, 2008.

[11] D. Lin and B. Xing, "Root uptake and phytotoxicity of ZnO nanoparticles," *Environmental Science & Technology*, vol. 42, no. 15, pp. 5580-5585, 2008.

[12] M. Kiser, H. Ryu, H. Jang, K. Hristovski, and P. Westerhoff, "Biosorption of nanoparticles to heterotrophic wastewater biomass," *Water Research*, vol. 44, no. 14, pp. 4105-4114, 2010.

[13] F. von der Kammer, S. Ottofuelling, T. Hofmann, "Assessment of the physico-chemical behavior of titanium dioxide nanoparticles in aquatic environments using multi-dimensional parameter testing," *Environmental Pollution*, vol. 158, no. 12, pp. 3472-3481.

[14] Xiaochen Zhang, Wen Li, and Zhuo Yang, "Toxicology of nanosized titanium dioxide: An update," *Archives of Toxicology* vol. 89, no. 12, pp. 2207-2217, 2015

[15] James C.K. Lai, Gayathri Ananthakrishnan, Sirisha Jandhyam, Vihas V. Dukhande, Alok Bhushan, Mugdha Gokhale, Christopher K. Daniels, and Solomon Leung, "Treatment of human hstrocytoma U87 cells with silicon dioxide nanoparticles lowers their survival and alters their expression of itochmondrial and cell signaling proteins," *International Journal of Nanomedicine*, vol 5, no.9, pp. 715-723, 2010

[16] James C.K. Lai, Maria B. Lai, Sirisha Jandhyam, Vikas V. Dukhande, Alok Bhushan, Christopher K. Daniels, and Solomon W. Leung, "Exposure to titanium dioxide and other metallic oxide nanoparticles induces cytotoxicity on human neural cells and fibroblasts," *International Journal of Nanomedicine*, vol. 3, no. 4, pp. 533-545, 2008.

[17] Jianhong Wu, Xue Wei, Shunchang Zhou, Fengli Lan, Lei Bi, Huibi Xu, Xiangliang Yang, and Fan-Dian Zeng, "Toxicity and penetration of TiO2 nanoparticles in hairless mice and porcine skin after subchronic dermal exposure," *Toxicol. Lett.*, vol. 191, no. 1, pp. 1-8, 2009.

[18] Jun-Jie Yin, Jun Liu, M. Ehrenshaft, J. Roberts, P. Fu, R. Mason, and Baozhong Zhao, "Phototoxicity of nano titanium dioxides in HaCaT keratinocytes-Generation of reactive oxygen species and cell damage," *Toxicol Appl Pharmacol.* vol. 263, no. 1, pp. 81-88, 2012

[19] S. Marquez-Ramirez, N. Delgado-Buenrostro, Y. Chirino, G. Iglesias, and R. Lopez-Marure, "Titanium dioxide nanoparticles inhibit proliferation and induce morphological changes and apoptosis in glial cells," *Toxicology.* vol. 302, no. 2-3, pp. 146-156, 2012.

[20] Florian Part, Gudrun Zecha, Tim Causon, Eva-Kathrin Sinner, and Marion Huber-Humer, "Current limitations and challenges in nanowaste detection, characterisation and monitoring," *Waste Management*, vol. 43, pp. 407-420, 2015.

[21] P. Westerhoff, g. Song, K. Hristovski, and M. Kiser, "Occurrence and removal of titanium at full scale wastewater treatment plants: implications for TiO2 nanomaterials," *Journal of Environmental Monitoring*, vol. 13, no. 5, pp. 1195-1203, May 2011

[22] A. Kunhikrishnan, H. Shon, N. Bolan, and I. El Saliby, "Sources, distribution, environmental fate, and ecological effects of nanomaterials in wastewater streams," *Environmental Sciences & Ecology*, vol. 45, no. 4, pp. 277-318, February 2015.

[23] G. Song, J. Wang, C. Chiu, and P. Westerhoff, "Biogenic nanoscale colloids in wastewater effluents," Environmenatl Science & Technology, vol. 44, no. 21, pp. 8216-8222.

[24] P. Westerhoff, A. Kiser, and K. Hristovski, "Nanomaterial removal and transformation during biological wastewater treatment," *Environmental Engineering Science*, vol. 30, no. 3, pp. 109-117, 2013.

[25] X. Zhou, B. Huang, T. Zhou, Y. Liu, and H. Shi, "Aggregation behavior of engineered nanoparticles and their impact on activated sludge in wastewater treatment," *Chemosphere*, vol. 119, pp. 568-576, 2015.

[26] L. Hou, K. Li, Y. Ding, Y. Li, J. Chen, X. Wu, and X. Li, "Removal of silver nanoparticles in simulated wastewater treatment processes and its impact on COD and NH4 reduction," *Chemosphere*, vol. 87, no. 3, pp. 248-252, 2012.